Hypercomplex Numbers

Translator's Notes

- The footnotes appear at the end of the book.
- The Russian original contains blocks of material set in fine print. In the English translation such material is preceded by the symbol ▷ and followed by the symbol ◁.

I.L. Kantor A.S. Solodovnikov

Hypercomplex Numbers

An Elementary Introduction to Algebras

Translated by A. Shenitzer

Springer-Verlag
New York Berlin Heidelberg
London Paris Tokyo

I.L. Kantor A.S. Solodovnikov

Translator
Professor Abe Shenitzer
Mathematics Department
York University
North York, Ontario
Canada M3J 1P3

Original Russian edition: *Giperkompleksnye Čisla*. Nauka, Moscow, 1973.

Mathematical Subject Classification Code (1980): 16-01, 17-A35, 16-03/01-01

Library of Congress Cataloging in Publication Data
Kantor, I.L. (Isai L'vovich)
[Giperkompleksnye chisla. English]
Hypercomplex numbers / I.L. Kantor, A.S .Solodovnikov; translated from the Russian by A. Shenitzer.
p. cm.
Translation of: Giperkompleksnye chisla.
ISBN 0-387-96980-2 (alk. paper)
1. Numbers, Complex. I. Solodovnikov. A.S. (Aleksandr Samuilovich) II. Title.
QA255.K3613 1989
512--dc20 89-6160

Printed on acid-free paper.

Camera-ready copy provided by the translator.
Printed and bound by Edwards Brothers, Inc., Ann Arbor, Michigan.
Printed in the United States of America.

9 8 7 6 5 4 3 2 1

ISBN 0-387-96980-2 Springer-Verlag New York Berlin Heidelberg
ISBN 3-540-96980-2 Springer-Verlag Berlin Heidelberg New York

Preface

This book deals with various systems of "numbers" that can be constructed by adding "imaginary units" to the real numbers. The complex numbers are a classical example of such a system.

One of the most important properties of the complex numbers is given by the identity

$$|zz'| = |z| \cdot |z'|. \tag{1}$$

It says, roughly, that the absolute value of a product is equal to the product of the absolute values of the factors. If we put $z = a_1 + a_2 i$, $z' = b_1 + b_2 i$, then we can rewrite (1) as

$$(a_1 b_1 - a_2 b_2)^2 + (a_1 b_2 + a_2 b_1)^2 = (a_1^2 + a_2^2)(b_1^2 + b_2^2).$$

The last identity states that "the product of a sum of two squares by a sum of two squares is a sum of two squares."

It is natural to ask if there are similar identities with more than two squares, and how all of them can be described. Already Euler had given an example of an identity with four squares. Later an identity with eight squares was found. But a complete solution of the problem was obtained only at the end of the 19th century.

It is substantially true that every identity with n squares is linked to formula (1), except that z and z' no longer denote complex numbers but more general "numbers"

$$a_1 + a_2 i + a_3 j + \cdots + a_n l,$$

where $i, j, \ldots, l$ are imaginary units. One of the main themes of this book is the establishing of the connection between identities with n squares and formula (1).

Another question we deal with at great length is *division* of hypercomplex numbers. The operations defined in each system of hypercomplex numbers are three of the four "arithmetic" operations, namely, addition, subtraction, and multiplication. The possibility of division depends on the system. Hypercomplex division systems (that is, systems with division) are few and far between. The real and complex numbers are, of course, examples of division systems. But there are others. The most remarkable of them are the *quaternions* and the *Cayley numbers*. The problem of finding *all* hypercomplex division systems is still open. Some variants of this problem will be considered in this book.

The first part familiarizes the reader with examples of hypercomplex numbers, including the quaternions and the Cayley numbers. The quaternions and the Cayley numbers (as well as some other hypercomplex systems) are division systems in which formula (1) holds. The third part explains the unique role of the complex numbers, the quaternions, and the Cayley numbers with respect to the questions we've raised. The second part is an elementary exposition of the fundamental concepts of linear algebra and is of an auxiliary nature.

The book is intended for students of science high schools and, less prescriptively, for all persons interested in mathematics. High school seniors should be able to read most of the material in the first two chapters, but may find that the reading of the rest of the book calls for rather strenuous efforts. Be that as it may, the reader need not worry about prerequisites.

Contents

Part I

Hypercomplex Numbers

Chapter 1

Complex Numbers

1.1 Introduction

In elementary algebra we consider, in addition to the real numbers, the larger system of *complex* numbers. What makes us study the complex numbers is the solution of quadratic equations. Specifically, certain quadratic equations, for example,

$$x^2 + 1 = 0, \tag{1.1}$$

cannot be solved if we restrict ourselves to the real numbers, that is, there is no real number a such that $a^2 = -1$.

The history of the complex numbers goes back to the 16th century. The Italian mathematicians Girolamo Cardano and Raffael Bombelli introduced the use of the symbol $\sqrt{-1}$, a formal solution of eq. (1.1), as well as the expression $b\sqrt{-1}$, a formal solution of the equation

$$x^2 + b^2 = 0.$$

Then the more general expression $a + b\sqrt{-1}$ can be regarded as a formal solution of the equation

$$(x - a)^2 + b^2 = 0. \tag{1.2}$$

Expressions of the form $a + b\sqrt{-1}$ came to be known first as imaginary numbers and then as *complex* numbers, and to be written as $a + bi$ (the use of i for $\sqrt{-1}$ goes back to the 18th century and is due to Euler). These numbers suffice to solve *all* quadratic equations. (We recall that if the discriminant of a quadratic equation is nonnegative, then its roots are real, and if the discriminant is negative, then its roots can be written in the form (1.2).)

Thus a complex number is an expression of the form

$$a + bi,$$

where a and b are real numbers and the symbol i is assigned the property $i^2 = -1$. We note that the complex numbers contain all real numbers (they are obtained by putting $b = 0$) as well as all "pure imaginary" numbers (obtained by putting $a = 0$.)

For the sake of brevity, we shall denote a complex number by the letter z and write

$$z = a + bi.$$

a is called the *real part* of z, and bi the *imaginary part* of z; i itself is called an "imaginary unit". The name "imaginary" should not be taken literally. It goes back to the time (the 16th and 17th centuries) when complex numbers were viewed as something unreal, and were surrounded by an aura of deep secrecy. In modern mathematics complex numbers are something quite natural and no more "imaginary" than the real numbers.

1.2 Operations on Complex Numbers

It is natural to define the operations of addition, subtraction, and multiplication of complex numbers as follows:

$$\begin{aligned} (a+bi)+(c+di) &= (a+c)+(b+d)i, \\ (a+bi)-(c+di) &= (a-c)+(b-d)i, \\ (a+bi)\times(c+di) &= ac+adi+bci+bdi^2 \\ &= (ac-bd)+(ad+bc)i \end{aligned}$$

(in the definition of multiplication we made use of the fact that $i^2 = -1$). Incidentally, if we put $b = 0$ in the definition of multiplication of complex numbers, then we obtain the rule

$$a(c + di) = ac + adi$$

for multiplying a real number by a complex number.

It is easy to verify that the laws governing the above operations are the same as the laws governing the corresponding operations on real numbers. Specifically, addition is commutative and associative:

$$z_1 + z_2 = z_2 + z_1, \qquad (z_1 + z_2) + z_3 = z_1 + (z_2 + z_3);$$

multiplication is also commutative and associative:

$$z_1 z_2 = z_2 z_1, \qquad (z_1 z_2) z_3 = z_1 (z_2 z_3);$$

and multiplication is distributive over addition:

$$z_1(z_2 + z_3) = z_1 z_2 + z_1 z_3. \tag{1.3}$$

To verify (1.3), say, we put

$$z_1 = a_1 + b_1 i,\ z_2 = a_2 + b_2 i,\ z_3 = a_3 + b_3 i.$$

Then

$$\begin{aligned} z_1(z_2 + z_3) &= (a_1 + b_1 i)[(a_2 + a_3) + (b_2 + b_3)i] \\ &= [a_1(a_2 + a_3) - b_1(b_2 + b_3)] + [a_1(b_2 + b_3) + b_1(a_2 + a_3)]i, \end{aligned}$$

$$\begin{aligned} z_1 z_2 + z_1 z_3 &= (a_1 + b_1 i)(a_2 + b_2 i) + (a_1 + b_1 i)(a_3 + b_3 i) \\ &= (a_1 a_2 - b_1 b_2 + a_1 a_3 - b_1 b_3) + (a_1 b_2 + b_1 a_2 + a_1 b_3 + b_1 a_3)i. \end{aligned}$$

It is easy to see that the outcomes of the two computations are the same.

1.3 The Operation of Conjugation

We consider in some detail certain further properties of the system of complex numbers.

With each complex number

$$z = a + bi$$

we associate its *complex conjugate*

$$\bar{z} = a - bi.$$

It is easy to see that

$$\overline{z_1 + z_2} = \bar{z}_1 + \bar{z}_2,$$

and

$$\overline{z_1 z_2} = \bar{z}_1 \bar{z}_2,$$

that is, the conjugate of a sum is the sum of the conjugates of the summands, and the conjugate of a product is the product of the conjugates of the factors. We leave it to the reader to check these formulas.

Note that

$$z + \bar{z} = 2a$$

and

$$z\bar{z} = a^2 + b^2,$$

that is, the sum and product of conjugate complex numbers are always real numbers.

1.4 The Absolute Value of a Complex Number: An Identity with Two Squares

The nonnegative real number $\sqrt{a^2+b^2}$ is called the *absolute value* of the complex number z and is denoted be $|z|$, that is,

$$|z| = \sqrt{a^2+b^2}.$$

We have

$$z\bar{z} = |z|^2.$$

The last equation implies a certain remarkable relation. Thus, let z_1 and z_2 be two complex numbers, then

$$|z_1 z_2|^2 = (z_1 z_2)(\overline{z_1 z_2}) = z_1 z_2 \bar{z}_1 \bar{z}_2 = z_1 \bar{z}_1 \cdot z_2 \bar{z}_2 = |z_1|^2 |\, z_2|^2,$$

so that

$$|z_1 z_2|^2 = |z_1|^2\, |z_2|^2, \tag{1.4}$$

and therefore

$$|z_1 z_2| = |z_1|\, |z_2|. \tag{1.5}$$

In other words, *the absolute value of a product is the product of the absolute values of the factors.* This is an extremely important property of the complex numbers; in chapter 16 we shall give it a special name (the property of *normability*). We shall now obtain a more detailed form of (1.4).

Put

$$z_1 = a_1 + b_1 i,\ z_2 = a_2 + b_2 i.$$

Then

$$z_1 z_2 = (a_1 a_2 - b_1 b_2) + (a_1 b_2 + a_2 b_1) i.$$

This means that we can write eq. (1.4) as

$$(a_1^2 + b_1^2)(a_2^2 + b_2^2) = (a_1 a_2 - b_1 b_2)^2 + (a_1 b_2 + a_2 b_1)^2.$$

This is an interesting identity. A somewhat vague formulation of this identity is that *the product of a sum of two squares by a sum of two squares is a sum of two squares.*

It is natural to ask if there are similar identities involving more squares. We shall see that this problem is anything but simple. It has occupied the minds of mathematicians for many years. It is one of the central issues considered in this book. We shall formulate it more precisely in chapter 3 and solve it in part 3.

1.5 Division of Complex Numbers

So far, we have said nothing about *division* of complex numbers. We discuss this next.

Let z' and z be two complex numbers and $z \neq 0$. By definition, the quotient z'/z is the solution of the equation

$$zx = z'. \tag{1.6}$$

Multiplying both sides of this equation by $\bar{z}$ we obtain $\bar{z}zx = \bar{z}z'$, so that

$$|z|^2 x = \bar{z}z'.$$

Multiplying both sides of the last equation by $1/|z|^2$ we have

$$x = (1/|z|^2)\bar{z}z'. \tag{1.7}$$

It is easy to check that the value of x in (1.7) satisfies equation (1.6).

We illustrate. Suppose that we wish to divide $z' = 5 - i$ by $z = 2 - 3i$. By formula (1.7),

$$\frac{z'}{z} = \frac{1}{2^2 + 3^2}(2 + 3i)(5 - i) = \frac{1}{13}(13 + 13i) = 1 + i.$$

Chapter 2

Alternate Arithmetics on the Numbers $a + bi$

2.1 Formulation of the Problem

We made the expressions $a + bi$ into a number system by introducing the following rules for their addition and multiplication:

$$(a + bi) + (c + di) = (a + c) + (b + d)i, \tag{2.1}$$

$$(a + bi)(c + di) = (ac - bd) + (ad + bc)i. \tag{2.2}$$

Formula (2.1) seems very natural. On the other hand, formula (2.2) does not inspire the same feeling. We shall now investigate the possibility of making the expressions $a+bi$ into a reasonable number system by retaining the addition rule (2.1) and replacing (2.2) by a multiplication rule that is, a priori, *arbitrary*.

The form of the new law depends largely on the properties we expect the new multiplication to have. It would be awkward if the new multiplication were given by the formula

$$(a + bi) \cdot (c + di) = ac^2 + bdi,$$

say. Indeed, putting $b = 0$, $d = 0$, we would then obtain the strange equality

$$a \cdot c = ac^2.$$

We shall expect the new multiplication to satisfy the following requirements:

1. Multiplication of a real number a, viewed as an element of the new system ($a = a + 0i$), by any number $z = b + ci$ must yield the same result as in the case of the complex numbers, that is,

   $$(a + 0i)(b + ci) = ab + aci,$$

   and

   $$(b + ci)(a + 0i) = ab + aci.$$

   In particular, this means that for real numbers the new multiplication must coincide with the usual multiplication:

   $$(a + 0i)(b + 0i) = ab + 0i.$$

   Since the analogous claim is true for addition ((2.1) implies that $(a + oi) + (b + 0i) = (a + b) + 0i$), it follows that the real numbers are included in the new system with their usual arithmetic.

2. The equality

   $$(az_1) \cdot (bz_2) = (ab) \cdot (z_1 z_2)$$

   must hold for all real a and b. For example, $(2i)(3i) = 6i^2$.

3. Multiplication must always be distributive over addition, that is, we must have

   $$z_1(z_2 + z_3) = z_1 z_2 + z_1 z_3,$$

   and

   $$(z_1 + z_2)z_3 = z_1 z_3 + z_2 z_3.$$

While these requirements do not determine the new law of multiplication completely, they imply a great deal. Thus,

$$\begin{aligned}(a + bi)(c + di) &= a(c + di) + (bi)(c + di)\\ &= ac + adi + bci + bdi^2.\end{aligned}$$

To determine the outcome completely, it remains to determine the value of i^2. In particular, if we put $i^2 = -1$, then we obtain the familiar multiplication of complex numbers. But this is certainly not the only possibility. After all, all that is required is that the product $i \cdot i$ belong to the number system under consideration, that is, that it be a number of the form $p + qi$. By assigning the values of p and q we will have completely determined the form of the multiplication law:

$$(a + bi)(c + di) = (ac + bdp) + (ad + bc + bdq)i. \tag{2.3}$$

Having thus defined the object of our study, we can dispense with the heuristic considerations that led us to formula (2.3) and say that *we investigate the system of numbers of the form* $a+bi$ *with addition rule* (2.1) *and multiplication rule* (2.3), *where* p *and* q *are two fixed real numbers* (that determine, so to say, the "arithmetic" of the given number system).

A close look at formula (2.3) indicates that the new multiplication is commutative ($z_1z_2 = z_2z_1$). This result is somewhat surprising, for it is not one of the three requirements imposed on the multiplication law. It is also true that our multiplication is associative. In fact,

$$\begin{aligned}[(a+bi)(c+di)](e+fi) &= [(ac+bdp)+(ad+bc+bdq)i](e+fi)\\ &= ((ac+bdp)e+(ad+bc+bdq)fp)\\ &+((ac+bdp)f+(ad+bc+bdq)e+(ad+bc+bdq)fq)i,\end{aligned}$$

$$\begin{aligned}(a+bi)[(c+di)(e+fi)] &= (a+bi)[(ce+dfp)+(cf+de+dfq)i]\\ &= (a(ce+dfp)+b(cf+de+dfq)p)\\ &+(a(cf+de+dfq)+b(ce+dfp)+b(cf+de+dfq)q)i.\end{aligned}$$

If we compare the results of the two computations, then we readily see that they are equal (to simplify the check, equal expressions appear on corresponding lines).

2.2 Reduction to Three Systems

The fact that formula (2.3) contains two arbitrary real numbers p and q may be taken as an indication that we have found an infinity of number systems. We are about to show that this is not at all so, and that each system can be reduced to one of the following three:

1. numbers $a+bi$ with $i^2=-1$ (the complex numbers);
2. numbers $a+bi$ with $i^2=1$ (the so-called double numbers);
3. numbers $a+bi$ with $i^2=0$ (the so called dual numbers).

The reduction process goes as follows. The equality $i^2=p+qi$ implies that $i^2-qi=p$, or that

$$(i-\frac{q}{2})^2 = p+\frac{q^2}{4}. \tag{2.4}$$

There are three possible cases:

case 1. $p + q^2/4$ is a negative number, that is, $p + q^2/4 = -k^2$, where k is a nonzero real number. Then

$$(i - \frac{q}{2})^2 = -k^2,$$

that is,

$$(-\frac{q}{2k} + \frac{1}{k}i)^2 = -1. \tag{2.5}$$

Denoting the number in parentheses by J, we have

$$J^2 = -1.$$

Since $i = q/2 + kJ$, it follows that every number $a + bi$ can be written in the form

$$a + bi = a + b(\frac{q}{2} + kJ) = (a + \frac{b}{2}q) + bkJ.$$

In other words, every number $a + bi$ can be written in the form $a' + b'J$, with $J^2 = -1$. *This means that we are actually dealing with the complex numbers.*

case 2. $p + q^2/4$ is a positive number, that is, $p + q^2/4 = k^2$. Then instead of (2.5) we have

$$(-\frac{q}{2k} + \frac{1}{k}i)^2 = 1.$$

Denoting the number in parentheses by E, we have

$$E^2 = 1.$$

It follows that every number $a+bi$ of our system can be written as $a'+b'E$, with $E^2 = 1$. The law of multiplication of these numbers is

$$(a' + b'E)(c' + d'E) = (a'c' + b'd') + (a'd' + b'c')E.$$

In other words, *for $p + q^2/4 > 0$ we obtain the system of double numbers.*

case 3. $p + q^2/4 = 0$. In this case, denoting by Ω the number $i - q/2$, we have

$$\Omega^2 = 0.$$

Every number $a + bi$ in our system can be written as $(a + bq/2) + b\Omega$, that is, in the form $\tilde{a} + \tilde{b}\Omega$. The law of multiplication takes the form

$$(\tilde{a} + \tilde{b}\Omega)(\tilde{c} + \tilde{d}\Omega) = \tilde{a}\tilde{c} + (\tilde{a}\tilde{d} + \tilde{b}\tilde{c})\Omega.$$

This means that we are dealing with the system of dual numbers.

In sum, we have shown that every system of numbers $a + bi$ with the operation rules (2.1) and (2.3) is actually one of the following three systems:

1. The complex numbers $a + bi, \quad i^2 = -1$;
2. The double numbers $a + bE, \quad E^2 = 1$;
3. The dual numbers $a + b\Omega, \quad \Omega^2 = 0$.

We have studied the properties of the complex numbers in considerable detail. Dual and double numbers are of lesser interest. Unlike the complex numbers, *dual and double numbers do not, in general, admit division*. We recall the meaning of the term "division": Given a law of multiplication, to divide z_1 by $z_2 (z_2 \neq 0)$ is to solve the equation

$$z_2 x = z_1 .$$

We show that in the system of double numbers it is not possible to divide, say, $z_1 = 1$ (that is, $1 + 0E$) by $z_2 = 1 + E$. Indeed, if the equation

$$(1 + E)x = 1 + 0E$$

were solvable, then it would follow that $(1 - E^2)x = 1 - E$, that is, $0 = 1 - E$, which is impossible. Similarly, in the system of dual numbers it is not possible to divide, say, 1 by Ω. Indeed, for every $x = a + b\Omega$ we have $x \cdot \Omega = a\,\Omega \neq 1$.

We are used to the notion that a key property of numbers is that we can add, subtract, multiply, and divide them. This being so, we might question the appropriateness of speaking of double and dual numbers. In mathematics, however, systems of "numbers" (similar to double and dual numbers) in which we can always add, subtract, and multiply, but not necessarily divide play an important role. Systems in which division can be carried out for all pairs $z_1, z_2 \neq 0$ are called *division systems.* In this book we shall be mainly concerned with division systems.

Chapter 3

Quaternions

3.1 Preliminaries

Our construction of the complex (as well as double and dual) numbers, suggests that we might go further, and consider numbers of the form

$$z = a + bi + cj,$$

where a, b, c are arbitrary real numbers and i and j are certain symbols. It is reasonable to adopt the following addition rule for these numbers:

$$(a + bi + cj) + (a' + b'i + c'j) = (a + a') + (b + b')i + (c + c')j.$$

The form of the multiplication rule requires some thought. Of course, we don't want a rule that would lead to awkward consequences, such as, say, the violation of the equality

$$(a + 0i + 0j)(b + 0i + 0j) = ab + 0i + 0j,$$

which states that for real numbers the new rule coincides with the usual multiplication of such numbers. Also, guided by the natural requirements stated in the previous section, we require that

1. The product of a real number $k = k+0i+0j$ by a number $z = a+bi+cj$ must be equal to $ka + kbi + kcj$;

2. The equality
$$(az_1)(bz_2) = (ab)(z_1 z_2)$$
must hold for arbitrary real numbers a, b; and

3. The distributive law

$$z_1(z_2 + z_3) = z_1 z_2 + z_1 z_3,$$

$$(z_1 + z_2)z_3 = z_1 z_3 + z_2 z_3,$$

must hold.

It is not difficult to invent a multiplication rule satisfying all these requirements. For example, we could adopt the rule

$$(a + bi + cj)(a' + b'i + c'j) = aa' + (ab' + ba')i + (ac' + ca')j.$$

While this multiplication rule is also commutative and associative ($z_1 z_2 = z_2 z_1$ and $(z_1 z_2)z_3 = z_1(z_2 z_3)$), it certainly does not imply the possibility of unrestricted division. For example, it is not possible to divide 1 by i, that is, the equation

$$(0 + 1i + 0j)x = 1 + 0i + 0j$$

has no solution.

This is not an accident. It is possible to show that for *every* multiplication rule satisfying 1, 2, and 3 there is at least one pair of numbers $z_1, z_2 (z_2 \neq 0)$ such that z_1 *cannot be divided by* z_2. In other words, it is impossible to make a division system out of the numbers $a + bi + cj$!

On the other hand, it is possible to make a division system out of the numbers

$$a + bi + cj + dk, \tag{3.1}$$

where k is an additional symbol. More precisely, it is possible to multiply the numbers (3.1) so that the requirements 1, 2, 3 hold and division, the inverse of multiplication, can always be carried out. The most interesting example of such a system are the *quaternions*.

3.2 The Definition of Quaternions

The quaternions are the numbers (3.1) with the addition rule

$$\begin{aligned}(a + bi + cj + dk) &+ (a' + b'i + c'j + d'k) \\ &= (a + a') + (b + b')i + (c + c')j + (d + d')k,\end{aligned}$$

and a rather special multiplication rule. To determine the multiplication rule it suffices to assign the values of the products of the numbers i, j, k

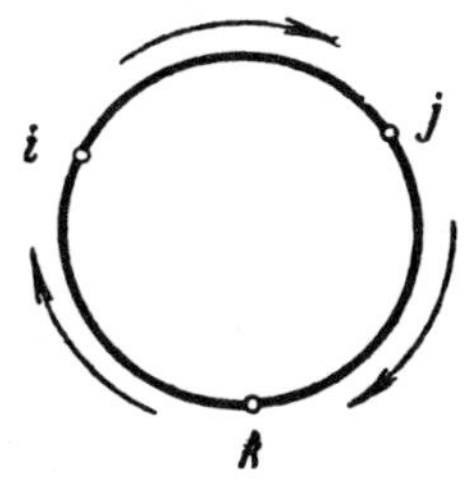

Figure 3.1.

taken two at a time:

$$\begin{gathered} i^2 = -1, \quad j^2 = -1, \quad k^2 = -1, \\ ij = k, \quad ji = -k, \\ jk = i, \quad kj = -i, \\ ki = j, \quad ik = -j. \end{gathered} \tag{3.2}$$

Figure 3.1 helps us remember this "multiplication table." In it, the numbers i, j, k are represented by dots placed clockwise on the circle. The product of two of these numbers is the third number or its negative, according as the orientation of the shortest circular arc joining the first factor to the second is clockwise or counterclockwise. We see that this multiplication rule is not commutative; the outcome depends on the order of the factors!

Our multiplication table and the requirements 1, 2, 3 enable us to multiply arbitrary quaternions. Thus, let

$$q = a + bi + cj + dk,$$

$$q' = a' + b'i + c'j + d'k.$$

By the rule for multiplying sums (implied by 3), we have

$$\begin{aligned} qq' &= aa' + a(b'i) + a(c'j) + a(d'k) + (bi)a' + (bi)(b'i) \\ &+ (bi)(c'j) + (bi)(d'k) + (cj)a' + (cj)(b'i) + (cj)(c'j) \\ &+ (cj)(d'k) + (dk)a' + (dk)(b'i) + (dk)(c'j) + (dk)(d'k). \end{aligned}$$

The terms with two of the three numbers i, j, k can be reduced using the requirements 1 and 2 and our multiplication table (for example, $(bi)(c'j) = bc'(ij) = bc'k$). The end result is

$$\begin{aligned} qq' &= (aa' - bb' - cc' - dd') + (ab' + ba' + cd' - dc')i \\ &+ (ac' + ca' + db' - bd')j + (ad' + da' + bc' - cb')k. \end{aligned} \tag{3.3}$$

3.3 Associativity of Multiplication of Quaternions

In spite of the fact that multiplication of quaternions is not commutative, computations involving quaternions are not as hard as might at first appear. What helps is that *multiplication of quaternions is associative*, that is,

$$(q_1 q_2)q_3 = q_1(q_2 q_3). \tag{3.4}$$

We verify this next.

Each of the quaternions $q_\alpha \, (\alpha = 1, 2, 3)$ is the sum of four terms ($q_\alpha = a_\alpha + b_\alpha j + c_\alpha j + d_\alpha k$). It follows that the left side of (3.4) is the sum of $4 \times 4 \times 4 = 64$ terms of the form

$$(u_1 u_2)u_3, \tag{3.5}$$

where u_1 is one of the summands in q_1, u_2 in q_2 and u_3 in q_3. Similarly, the right side of (3.4) is the sum of 64 terms

$$u_1(u_2 u_3). \tag{3.6}$$

If we can show that each of the terms (3.5) is equal to some term (3.6), then we'll have proved (3.4).

Thus, to verify (3.4) it suffices to verify it for the special case when q_1, q_2, q_3 are any three of the four quaternions $a, b\,i, cj, d\,k$. Since we can pull out numerical coefficients, we need only verify (3.4) for the four quaternions $1, i, j, k$; for example, instead of showing that $((b\,i)(cj))(b'i) = (b\,i)((cj)(b'i))$, it suffices to show that $(ij)i = i(ji)$.

If one of the quaternions q_1, q_2, q_3 is 1, then (3.4) is obviously true. Thus it suffices to verify (3.4) when q_1, q_2, q_3 are any of the quaternions i, j, k. There are 27 such equalities. Some of them are

$$(ii)i = i(ii), \quad (ii)j = i(ij), \quad (ij)i = i(ji), \quad (ij)k = i(jk).$$

Using table (3.2) we can easily check all 27 equalities. This proves the associativity of the multiplication of quaternions.

We shall see that the system of quaternions resembles the system of complex numbers in many very important respects. We have just checked that the multiplication of quaternions is associative. But the similarities between the two systems are far greater. As already indicated, the quaternions admit division. Then there is the possibility of defining the absolute value of a quaternion so that "the absolute value of a product is the product of the absolute values" of the factors.

What is behind these similarities is the possibility of defining on the quaternions an operation of conjugation whose properties are analogous to those of the conjugation of complex numbers.

3.4 Conjugation of Quaternions

By analogy with the complex numbers, we make the following definition.

By the *conjugate* of the quaternion

$$q = a + bi + cj + dk,$$

we mean the quaternion

$$\bar{q} = a - bi - cj - dk. \tag{3.7}$$

It is clear that the sum of conjugate quaternions is a real number. The quaternion multiplication rule 3 implies that the product $q\bar{q}$ is also real. In fact,

$$(a + bi + cj + dk)(a - bi - cj - dk) = a^2 + b^2 + c^2 + d^2. \tag{3.8}$$

Continuing the analogy with the complex numbers, we call the nonnegative number

$$\sqrt{a^2 + b^2 + c^2 + d^2}$$

the *absolute value* of the quaternion q and denote it by $|q|$. Then (3.8) can be written as

$$q\bar{q} = |q|^2.$$

This formula is the same as the one for complex numbers.

Remark. If q' is a "pure imaginary" quaternion, that is, if $q' = bi + cj + dk$, then

$$q'^2 = -(b^2 + c^2 + d^2) \leq 0.$$

Conversely, if the square of a quaternion is real and less than or equal to zero, then that quaternion is pure imaginary. (In fact, if $q = a+bi+cj+dk$, then $q^2 = (a+q')(a+q') = a^2 + q'^2 + 2aq' = a^2 - b^2 - c^2 - d^2 + 2aq'$. If the last expression were a real number and $a \neq 0$, then $q' = 0$. But then $q = a$ and q^2 is not ≤ 0.) It follows that quaternions of the form $bi + cj + dk$ — and only such quaternions — can be characterized by the condition that their squares are real numbers ≤ 0. With this in mind, we can give the following alternate description of the operation of conjugation: *let q be a quaternion and let $q = a + q'$ be its unique representation such that the square of the quaternion q' is real and ≤ 0. Then $\bar{q} = a - q'$.* This remark will be used in chapter 17.

Direct verification shows that conjugation has the properties

$$\overline{q_1 + q_2} = \bar{q}_1 + \bar{q}_2 \tag{3.9}$$

(the conjugate of a sum is the sum of the conjugates of the summands), and

$$\overline{q_1 q_2} = \bar{q}_2 \bar{q}_1 \tag{3.10}$$

(the conjugate of a product is the product of the conjugates of the factors in reverse order). The reader will recall that analogous equalities hold for complex numbers. The one difference is that whereas for complex numbers we can write $\bar{z}_1 \bar{z}_2$ for $\bar{z}_2 \bar{z}_1$ (multiplication of complex numbers is commutative), the quaternion products $\bar{q}_2 \bar{q}_1$ and $\bar{q}_1 \bar{q}_2$ are, in general, different.

To verify (3.10) it suffices to check it in those cases in which q_1 and q_2 are any two of the three quaternions i, j, k. Then the verification is easily accomplished by using the table (3.2). For example,

$$\overline{ii} = \overline{-1} = -1, \quad \text{and} \quad \bar{i}\bar{i} = (-i)(-i) = i^2 = -1,$$
$$\overline{ij} = \bar{k} = -k, \quad \text{and} \quad \bar{j}\bar{i} = (-j)(-i) = ji = -k,$$

and so on.

3.5 The Quaternions as a Division System

There is a basic difference between division of quaternions and division of complex numbers. The reader will recall that for complex numbers the quotient of z_1 by z_2 is the solution of the equation $z_2 x = z_1$. But multiplication of quaternions is noncommutative, and so it is necessary to consider not one but two equations:

$$q_2 x = q_1 \tag{3.11}$$

and

$$x q_2 = q_1. \tag{3.11'}$$

We call the solution of the first of these equations the *left quotient* of q_1 by q_2 and denote it by x_l. Similarly, we call the solution of the second equation the *right quotient* of q_1 by q_2 and denote it by x_r. (Of course, for complex numbers the two inverses coincide.)

To solve the equations (3.11) and (3.11′) we use the approach used earlier in connection with complex numbers. We multiply both sides of (3.11) on the left by $\bar{q}_2$ and then by $\frac{1}{|q_2|^2}$. The result is

$$x = \frac{1}{|q_2|^2} \bar{q}_2 q_1.$$

Substitution in (3.11) shows that this expression is a solution. Hence

$$x_l = \frac{1}{|q_2|^2} \bar{q}_2 q_1.$$

Similarly,

$$x_r = \frac{1}{|q_2|^2}\, q_1 \bar{q}_2.$$

By way of an example, we compute the left and right quotient of k by $1 + i + k$:

$$x_l = \frac{1}{3}(1 - i - k)k = \frac{1}{3}(k + j + 1),$$

$$x_r = \frac{1}{3}k(1 - i - k) = \frac{1}{3}(k - j + 1).$$

In sum, we have established the two most important properties of the system of quaternions:

1. Multiplication of quaternions is associative;

2. The quaternions are a division system.

3.6 Absolute Value of a Product

One other important property of quaternions is that *the absolute value of a product is the product of the absolute values of the factors.* Its proof is the same as for complex numbers. It makes use of the equality $\overline{q_1 q_2} = \bar{q}_2 \bar{q}_1$ and the associativity of quaternion multiplication. The proof follows.

$$|q_1 q_2|^2 = (q_1 q_2)(\overline{q_1 q_2}) = (q_1 q_2)(\bar{q}_2 \bar{q}_1) = q_1 (q_2 \bar{q}_2) \bar{q}_1 = |q_1|^2\, |q_2|^2.$$

3.7 The Four-Square Identity. General Formulation of the Problem of the Sum of Squares

The equality

$$|q_1 q_2|^2 = |q_1|^2\, |q_2|^2, \tag{3.12}$$

spelled out in detail, leads to an interesting identity. Thus, let

$$q_1 = a + bi + cj + dk, \quad q_2 = a' + b'i + c'j + d'k.$$

Then $q_1 q_2$ is the expression on the right side of (3.3). If we reverse sides in (3.12), then we can write it as

$$\begin{aligned}(a^2 + b^2 + c^2 + d^2)&(a'^2 + b'^2 + c'^2 + d'^2) \\ &= (aa' - bb' - cc' - dd')^2 + (ab' + ba' + cd' - dc')^2 \\ &+ (ac' + ca' + db' - bd')^2 + (ad' + da' + bc' - cb')^2.\end{aligned} \tag{3.13}$$

We recall that for complex numbers the equality $|z_1 z_2|^2 = |z_1|^2|z_2|^2$ yields the analogous identity

$$(a^2 + b^2)(a'^2 + b'^2) = (aa' - bb')^2 + (ab' + ba')^2, \qquad (3.14)$$

which we said is to the effect that the product of the sum of two squares is again a sum of two squares. Similarly, (3.13) may be said to assert that *the product of the sum of four squares by a sum of four squares is again a sum of four squares.*

The above identities suggest the following problem: For what values of n are there identities stating that "the product of a sum of n squares by a sum of n squares is a sum of n squares"?

For $n = 1$ we have the immediate positive answer

$$a^2b^2 = (ab)^2.$$

As we know, for $n = 2$ and $n = 4$ the answer, while not at all obvious, is again positive. But what about $n = 3, 5, 6$ and so on? As noted earlier, for a long time the problem was not fully answered. We owe a complete answer to the German mathematician A. Hurwitz who showed in 1898 that *identities of the required kind exist for* $n = 1, 2, 4, 8$ *and for no other values of* n.

To avoid possible misunderstandings concerning "the problem of the sum of squares" we restate it with greater precision.

Let $a_1, a_2, \ldots, a_n$ and $b_1, b_2, \ldots, b_n$ be two strings of letters. By a *bilinear form* in these letters we mean a sum such that each summand is a product of a letter in the first string and a letter in the second string. For example, the expression

$$a_1b_1 + 8a_1b_2 - 2a_3b_5 + 3a_3b_n$$

is a bilinear form. "The problem of the sum of squares" can be stated precisely as follows: *For what values of n one can find n bilinear forms $\Phi_1, \Phi_2, \ldots, \Phi_n$ such that*

$$(a_1^2 + a_2^2 + \ldots + a_n^2)(b_1^2 + b_2^2 + \ldots + b_n^2) = \Phi_1^2 + \Phi_2^2 + \ldots + \Phi_n^2. \; (!)$$

Clearly, the identities (3.13) and (3.14), which we found in connection with our study of quaternions and complex numbers, are of the required type. After a minor change of notation, the identities in question can be rewritten as

$$(a_1^2 + a_2^2)(b_1^2 + b_2^2) = (a_1b_1 - a_2b_2)^2 + (a_1b_2 + a_2b_1)^2,$$

and

$$(a_1^2 + a_2^2 + a_3^2 + a_4^2)(b_1^2 + b_2^2 + b_3^2 + b_4^2)$$
$$= (a_1b_1 - a_2b_2 - a_3b_3 - a_4b_4)^2 + (a_1b_2 + a_2b_1 + a_3b_4 - a_4b_3)^2$$
$$+(a_1b_3 + a_3b_1 + a_4b_2 - a_2b_4)^2 + (a_1b_4 + a_4b_1 + a_2b_3 - a_3b_2)^2.$$

In chapter 6 we shall consider the division system known as Cayley numbers and obtain from them an identity (!) for $n = 8$. In this way we shall have obtained (!) for $n = 1, 2, 4, 8$. In Part 3, after discussing all the necessary preliminaries, we shall prove the previously mentioned Hurwitz theorem which asserts the impossibility of (!) for other values of n.

Chapter 4

Quaternions and Vector Algebra

▷ The discovery of quaternions in the middle of the 19th century provided the impulse for a variety of researches in mathematics and physics. In particular, quaternions gave rise to *vector algebra*, one of the most fruitful areas of mathematics. In this section we shall describe the connection between the calculus of quaternions and the operations on vectors in 3-space.

4.1 The Number and Vector Parts of a Quaternion

We recall certain issues which the reader learned in geometry. Consider a rectangular coordinate system in space with unit vectors i, j, k on the coordinate axes (Figure 4.1). Then any sum of the form $bi + cj + dk$, where b, c, d are real numbers, represents a vector joining the origin O of the coordinate system to the point M with coordinates b, c, d.

Now consider the quaternions. We may regard each quaternion

$$q = a + bi + cj + dk$$

as a formal sum of the real number a and the vector $bi + cj + dk$. We shall call a the *number* (or *real)* part of q, and $bi + cj + dk$ its *vector* (or *imaginary)* part.

Now consider the vector quaternions

$$q_1 = b_1 i + c_1 j + d_1 k, \quad q_2 = b_2 i + c_2 j + d_2 k,$$

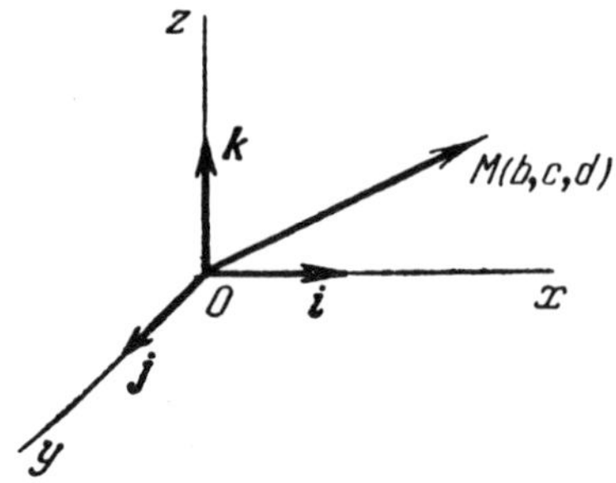

Figure 4.1.

and their product

$$\begin{aligned} q_1 q_2 &= -(b_1 b_2 + c_1 c_2 + d_1 d_2) + (c_1 d_2 - d_1 c_2) i \\ &\quad +(d_1 b_2 - b_1 d_2) j (b_1 c_2 - c_1 b_2) k. \end{aligned} \tag{4.1}$$

We have:

$$\text{Real part of } q_1 q_2 = -(b_1 b_2 + c_1 c_2 + d_1 d_2). \tag{4.2}$$

Imaginary part of

$$q_1 q_2 = (c_1 d_2 - d_1 c_2) i + (d_1 b_2 - b_1 d_2) j + (b_1 c_2 - c_1 b_2) k. \tag{4.3}$$

4.2 Scalar Product of Vectors

Each of the expressions (4.2) and (4.3) has a definite geometric sense. We are about to show that the sum $b_1 b_2 + c_1 c_2 + d_1 d_2$ is equal to $|q_1|\,|q_2| \cos\varphi$, that is, the product of the lengths (or absolute values) of the vectors q_1 and q_2 by the cosine of the angle between them. Such products turn up in mathematics with extraordinary frequency. We refer to such a product as "the scalar product of the vectors q_1 and q_2." (We emphasize that the scalar product is a number and not a vector.) We denote it by (q_1, q_2). Thus, by definition,

$$(q_1, q_2) = |q_1|\,|q_2| \cos\varphi.$$

We prove that

$$(q_1, q_2) = b_1 b_2 + c_1 c_2 + d_1 d_2. \tag{4.4}$$

Consider the triangle in Figure 4.2 determined by the vectors q_1 and q_2. One of its vertices is at the origin. The remaining vertices are the points

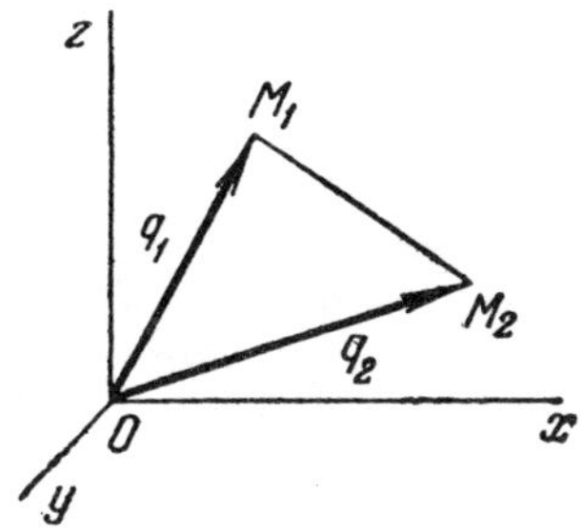

Figure 4.2.

M_1, M_2 (the endpoints of the vectors q_1, q_2) with coordinates b_1, c_1, d_1 and b_2, c_2, d_2, respectively. We have

$$OM_1^2 = b_1^2 + c_1^2 + d_1^2,$$

$$OM_2^2 = b_2^2 + c_2^2 + d_2^2,$$

$$M_1M_2^2 = (b_1 - b_2)^2 + (c_1 - c_2)^2 + (d_1 - d_2)^2,$$

so that

$$M_1M_2^2 = OM_1^2 + OM_2^2 - 2(b_1b_2 + c_1c_2 + d_1d_2).$$

By the law of cosines,

$$M_1M_2^2 = OM_1^2 + OM_2^2 - 2OM_1 \cdot OM_2 \cdot \cos\varphi,$$

where φ is the angle at O (the angle between the vectors q_1 and q_2). Equating the expressions for $M_1M_2^2$ we see that

$$OM_1 \cdot OM_2 \cdot \cos\varphi = b_1b_2 + c_1c_2 + d_1d_2,$$

which was to be proved.

Thus the real part of the product of the vector quaternions q_1, q_2 is the negative of their scalar product.

We observe that if the nonzero vectors q_1 and q_2 are perpendicular, then their scalar product is zero ($\varphi = \frac{\pi}{2}$, $\cos\varphi = 0$). But then the real part of the product is zero and q_1q_2 is a "pure" vector. Of course, the converse is also true: if q_1q_2 is a pure vector, then the scalar product of q_1 and q_2 is zero. But then, assuming that $q_1, q_2 \neq 0$, that is, that φ is defined, it follows that $\cos\varphi = 0$ and q_1, q_2 are perpendicular. Also, if q_1 and q_2 are perpendicular, then $q_1q_2 = -q_2q_1$. This follows readily from formula (4.1) if we bear in mind that the real part of q_1q_2 is zero.

4.3 Cross Product of Vectors

The geometric interpretation of the vector part of the product q_1q_2, that is, of the right side of (4.3), is more difficult. We call it the *cross product* of the vectors q_1 and q_2 and denote it by $[q_1, q_2]$. Thus $[q_1, q_2] = (c_1d_2 - d_1c_2)i + (d_1b_2 - b_1d_2)j + (b_1c_2 - c_1b_2)k$.

It turns out that the vector $[q_1, q_2]$ is perpendicular to each of the vectors q_1 and q_2 and that its length is equal to $|q_1|\,|q_2| \sin\varphi$, that is, the area S of the parallelogram on the vectors q_1 and q_2.

We know that in order to prove the perpendicularity of the vectors $[q_1, q_2]$ and q_1 it suffices to show that the real part of the product of these quaternions is zero or, what amounts to the same thing, that their product is a pure vector. Now (4.1) and (4.4) imply that $[q_1, q_2] = q_1q_2 + (q_1, q_2)$, so that

$$\begin{aligned} q_1[q_1, q_2] &= q_1(q_1q_2 + (q_1, q_2)) \\ &= q_1^2q_2 + (q_1, q_2)q_1 = -|q_1|^2q_2 + (q_1, q_2)q_1. \end{aligned}$$

(Note that in the computation process we replaced q_1^2 by $-|q_1|^2$. This is justified by formula (4.1), which implies that $q_1^2 = -(b_1^2 + c_1^2 + d_1^2) + 0i + 0j + 0k = -|q_1|^2$.) The expression on the right is a sum of two vectors, and thus a vector. A similar argument establishes the perpendicularity of the vectors $[q_1, q_2]$ and q_2.

It remains to compute the length of the vector $[q_1, q_2]$. Its square is equal to

$$(c_1d_2 - d_1c_2)^2 + (d_1b_2 - b_1d_2)^2 + (b_1c_2 - c_1b_2)^2,$$

which can be rewritten as

$$(b_1^2 + c_1^2 + d_1^2)(b_2^2 + c_2^2 + d_2^2) - (b_1b_2 + c_1c_2 + d_1d_2)^2.$$

The last expression is $|q_1|^2\,|q_2|^2 - (q_1, q_2)^2$ or, bearing in mind the definition of the scalar product,

$$|q_1|^2\,|q_2|^2 - |q_1|^2\,|q_2|^2 \cos^2\varphi, \quad \text{i.e., } |q_1|^2\,|q_2|^2 \sin^2\varphi.$$

Thus, as was to be shown, the square of the length of the vector $[q_1, q_2]$ is equal to $|q_1|^2\,|q_2|^2 \sin^2\varphi$. , that is, S^2.

The properties of the vector $[q_1, q_2]$ just established (the fact that it is perpendicular to both q_1 and q_2 and that its length is S) do not determine it uniquely. In fact, there are *just two* such vectors, and they are oppositely directed (Figure 4.3). The description of the vector $[q_1, q_2]$ is completed by the statement that the orientation of the triple $q_1, q_2, [q_1, q_2]$ in space is the same as that of the triple i, j, k. By this we mean the following: If we

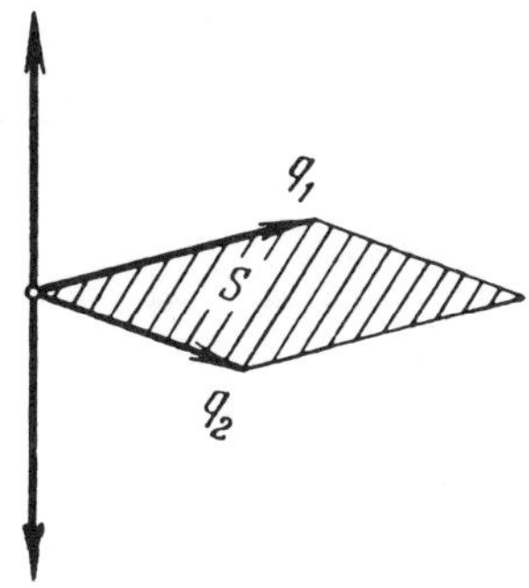

Figure 4.3.

look at the plane of the vectors q_1 and q_2 from the tip of the vector $[q_1, q_2]$, then we see that the orientation of the smallest rotation from q_1 to q_2 is the same as the orientation of the smallest rotation from i to j, determined by viewing the plane of i and j from the tip of k (Figure 4.4).

In sum, for the multiplication of pure vector quaternions we have the formula

$$q_1 q_2 = -(q_1, q_2) + [q_1, q_2].$$

Here (q_1, q_2) is the scalar product of the vectors q_1 and q_2, and $[q_1, q_2]$ is their cross product. We see that the scalar and cross products are "fragments" of quaternion multiplication.

The operations of scalar and cross product (together with vector addition and multiplication of vectors by scalars) are the basis of vector algebra — a branch of mathematics with numerous applications in mathematics and in physics (especially mechanics). The reader may be familiar with some of these applications (work is the scalar product of the force vector by the displacement vector, and so on). It should be noted that a clear presentation of vector algebra appeared much later than the first papers on the theory of quaternions (the papers of the English mathematician Hamilton, the founder of quaternion theory, appeared in the 1850s, whereas the basic aspects of vector algebra were formulated by the American mathematician and physicist Gibbs in the 1880s).

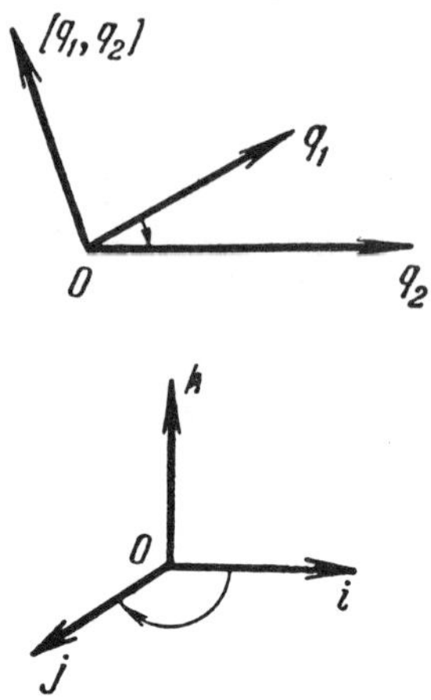

Figure 4.4.

4.4 The Geometric Interpretation of the Multiplication of a Quaternion by a Pure Vector Quaternion

Owing to the fact that quaternion multiplication involves the scalar and cross product of vectors, quaternions are a remarkable tool for the solution of certain problems in mechanics and geometry. Below we state a very difficult problem which can be solved by means of quaternions in a manner that is at once very simple and beautiful. Before we can do this, however, we must explain the geometric significance of the multiplication of a quaternion by a pure vector quaternion.

Let

$$q = a + bi + cj + dk$$

be a quaternion whose absolute value is 1. Then

$$a^2 + b^2 + c^2 + d^2 = 1.$$

Put

$$q = a + q',$$

where q' is the vector $bi + cj + dk$. Since $|a^2| + |q'|^2 = 1$, there exists a unique angle $\varphi, 0^0 \leq \varphi \leq 180^0$, such that

$$a = \cos\varphi, \quad |q'| = \sin\varphi.$$

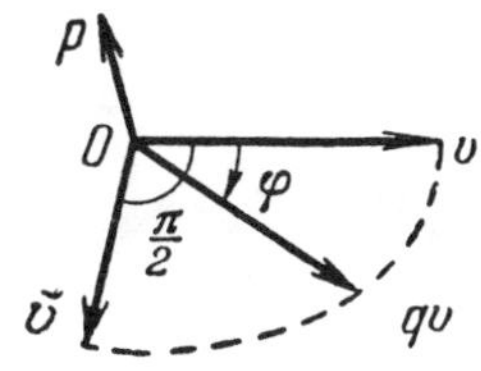

Figure 4.5.

Clearly, $q' = |q'|\,p$, where p is a vector of unit length. Hence

$$q = \cos\varphi + p\sin\varphi.$$

We emphasize that every quaternion q of absolute value 1 can be so represented (with p a vector of unit length) and that this representation is unique.

We multiply the quaternion q by a vector quaternion v that is *perpendicular to p* but otherwise arbitrary. We have

$$qv = (\cos\varphi + p\sin\varphi)v = v\cos\varphi + pv\sin\varphi.$$

Since q and v are perpendicular, the real part of pv is zero and its vector part is $[p, v]$, that is, a vector of length $|p|\cdot|v|\cdot\sin\frac{\pi}{2} = |v|$, perpendicular to p and v and oriented with respect to p and v in the same way as the vector k with respect to i and j. Denote this vector by $\tilde{v}$. Then we can say that $\tilde{v}$ is the result of rotating v through $\pi/2$ about p. To avoid any ambiguity we stipulate that the orientation of the rotation about p is to be the same as the orientation of the smallest rotation from i to j about k.[1] We have

$$qv = v\cos\varphi + \tilde{v}\sin\varphi.$$

A glance at Figure 4.5 shows that the vector qv is obtained from v by a rotation through φ about the vector p.

Thus, *if p is a vector of length* 1 *and v is any vector perpendicular to p, then by multiplying v on the left by the quaternion $q = \cos\varphi + p\sin\varphi$ we rotate it about p through the angle φ.*

Up to a point, this fact may be regarded as the geometric sense of multiplication (on the left) by q. What is disappointing is that the vector v is not arbitrary but perpendicular to p.

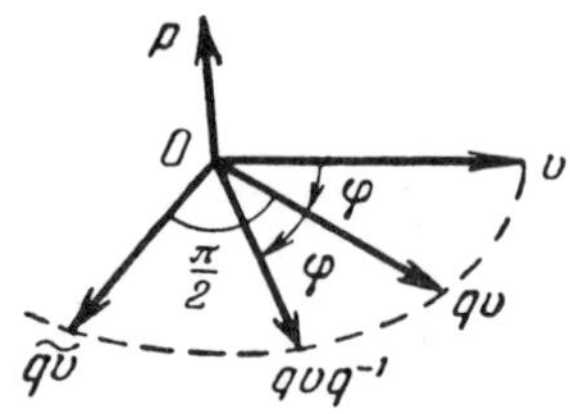

Figure 4.6.

4.5 Representation of an Arbitrary Rotation in Space by Means of Quaternions

By using a more complicated action on v we can obtain a quaternion representation of a rotation about p of an *arbitrary* vector v. Specifically, instead of the product qv we must consider the more complex expression

$$qvq^{-1}.$$

Here q^{-1} is the inverse of the quaternion q, that is, $qq^{-1} = 1$. It is easy to see that

$$q^{-1} = \cos\varphi - p\sin\varphi$$

(in fact,$(\cos\varphi + p\sin\varphi)(\cos\varphi - p\sin\varphi) = \cos^2\varphi - p^2\sin^2\varphi = \cos^2\varphi + \sin^2\varphi = 1$).

We shall show that *the vector qvq^{-1} is the result of rotating the vector v about the vector p through 2φ.*

First assume that v is perpendicular to p. Then

$$qvq^{-1} = qv(\cos\varphi - p\sin\varphi) = qv\cos\varphi - (qv)p\sin\varphi.$$

We know that qv is again a vector perpendicular to p. Hence $(qv)p = -p(qv)$. Earlier we saw that the quaternion $p(qv)$ is the vector obtained by rotating qv about the vector p through $\pi/2$ (Figure 4.6). As before, we denote it by $\widetilde{qv}$. Thus

$$qvq^{-1} = qv\cos\varphi + \widetilde{qv}\sin\varphi.$$

The expression on the right is the vector obtained by rotating qv about p through the angle φ. If we bear in mind that the vector qv is obtained from

v by the same rotation, then it is clear that qvq^{-1} is the result of rotating v about p through the angle 2φ.

Before considering the general case we observe that if the vector v is a multiple of $p, v = \lambda p$, then clearly, $qv = vq$ and

$$qvq^{-1} = vqq^{-1} = v.$$

Now let v be an arbitrary vector. We decompose it into two components, $v = v_1 + v_2$, where v_1 is a vector perpendicular to p and v_2 is proportional to p. Then

$$qvq^{-1} = qv_1q^{-1} + qv_2q^{-1} = qv_1q^{-1} + v_2.$$

We see that the component v_1 is rotated about p through the angle 2φ and the component v_2 remains unchanged. But then v is rotated about p through the angle 2φ.

We have shown that *the rotation about p through the angle 2φ takes the vector v into the vector qvq^{-1}*, where

$$q = \cos\varphi + p\sin\varphi.$$

With this in mind, we say that the indicated rotation *corresponds* to the quaternion q.

4.6 The Problem of "Composition" of Rotations

In the beginning of section 4.4 we promised to illustrate the application of quaternions by using them to solve a difficult geometric problem. We do this next.

Consider a rotation through an angle $2\varphi_1$ about an axis determined by a unit vector p_1. This rotation is followed by a rotation through an angle $2\varphi_2$ about an axis determined by a unit vector p_2. We are required to find the axis and angle of the resultant rotation.

We know that the first rotation takes any vector v into the vector $v_1 = q_1vq_1^{-1}$, where $q_1 = \cos\varphi_1 + p_1\sin\varphi_1$. The second rotation takes v_1 into

$$v_2 = q_2v_1q_2^{-1} = q_2(q_1vq_1^{-1})q_2^{-1} = (q_2q_1)v(q_2q_1)^{-1}$$

(here we have used the equality $(q_2q_1)^{-1} = q_1^{-1}q_2^{-1}$, implied by the equality $(q_2q_1)(q_1^{-1}q_2^{-1}) = 1$). Thus, the successive application of the two rotations takes the vector v into the vector

$$v_2 = (q_2q_1)v(q_2q_1)^{-1}.$$

Put differently, *the result of successive application of two rotations corresponding to the quaternions q_1 and q_2 is the rotation corresponding to the quaternion q_2q_1.*

Since we have a rule for multiplying quaternions, we can easily compute q_2q_1. Then we put it in the form

$$q_2q_1 = \cos\psi + p\sin\psi \tag{4.5}$$

with p a unit vector. The resultant rotation is a rotation about p through the angle 2ψ. It is certainly true that by using quaternions we have solved our problem with ease!

We illustrate by means of an example. Suppose the first rotation is about the x-axis through the angle $\pi/2$ and the second rotation is about the y-axis through the same angle. The quaternion corresponding to the first rotation is $q_1 = \cos\frac{\pi}{4} + i\sin\frac{\pi}{4} = \frac{\sqrt{2}}{2}(1+i)$, and the quaternion corresponding to the second rotation is $q_2 = \frac{\sqrt{2}}{2}(1+j)$. Hence

$$q_2q_1 = \frac{1}{2}(1+j)(1+i) = \frac{1}{2}(1+i+j-k).$$

To put this quaternion in the form (4.5) we note that its real part is $\frac{1}{2} = \cos\pi/3$. Hence

$$q_2q_1 = \cos\frac{\pi}{3} + [\frac{1}{\sqrt{3}}(i+j-k]\sin\frac{\pi}{3}$$

It follows that the resultant is the rotation about the vector $p = \frac{1}{\sqrt{3}}(i+j-k)$ through the angle $2\pi/3$. ◁

Chapter 5

Hypercomplex Numbers

5.1 Definition of a Hypercomplex Number System

Complex, double, and dual numbers, as well as quaternions, are all instances of *hypercomplex number systems*. Now that the reader is familiar with the simplest examples of such systems he will find it easier to appreciate their more general definition.

Consider expressions of the form

$$a_0 + a_1\boldsymbol{i}_1 + a_2\boldsymbol{i}_2 + \ldots + a_n\boldsymbol{i}_n, \tag{5.1}$$

where n is a fixed integer, $a_0, a_1, a_2, \ldots a_n$ are arbitrary real numbers and $\boldsymbol{i}_1, \boldsymbol{i}_2, \ldots, \boldsymbol{i}_n$ are certain symbols (that we shall sometimes refer to as "imaginary units"). By definition

$$a_0 + a_1\boldsymbol{i}_1 + \ldots + a_n\boldsymbol{i}_n = b_0 + b_1\boldsymbol{i}_1 + \ldots + b_n\boldsymbol{i}_n,$$

if and only if

$$a_0 = b_0, \quad a_1 = b_1, \quad \ldots, \quad a_n = b_n.$$

For the sake of brevity, we shall denote the expressions (5.1) by means of single, boldface letters $\boldsymbol{a}, \boldsymbol{b}, \boldsymbol{c}, \boldsymbol{u}, \boldsymbol{v}, \boldsymbol{w}$, etc. . An exception to this rule is the use of a_0 for expressions of the form

$$a_0 + 0\boldsymbol{i}_1 + 0\boldsymbol{i}_2 + \ldots + 0\boldsymbol{i}_n.$$

We shall add, subtract, and multiply the expressions (5.1). Addition and subtraction are defined by the formulas

$$(a_0 + a_1\boldsymbol{i}_1 + \ldots + a_n\boldsymbol{i}_n) + (b_0 + b_1\boldsymbol{i}_1 + \ldots + b_n\boldsymbol{i}_n) =$$
$$= (a_0 + b_0) + (a_1 + b_1)\boldsymbol{i}_1 + \ldots + (a_n + b_n)\boldsymbol{i}_n,$$

$$(a_0 + a_1\boldsymbol{i}_1 + \ldots + a_n\boldsymbol{i}_n) - (b_0 + b_1\boldsymbol{i}_1 + \ldots + b_n\boldsymbol{i}_n)$$
$$= (a_0 - b_0) + (a_1 - b_1)\boldsymbol{i}_1 + \ldots + (a_n - b_n)\boldsymbol{i}_n.$$

Multiplication is defined as follows.

We prescribe a "multiplication table," that is, we assign to each "product" $\boldsymbol{i}_\alpha\boldsymbol{i}_\beta$ a "value" of the form (5.1); here α and β are integers from 1 to n. (Clearly, the number of such products is $n \cdot n = n^2$.) In other words,

$$\boldsymbol{i}_\alpha\boldsymbol{i}_\beta = p_0 + p_1\boldsymbol{i}_1 + p_2\boldsymbol{i}_2 + \ldots + p_n\boldsymbol{i}_n, \tag{5.2}$$

where the choice of the real numbers $p_0, p_1, \ldots, p_n$ is uniquely determined by the choice of the subscripts α, β. To stress the dependence of the coefficients in (5.2) on the choice of α, β we write $p_{\alpha\beta,i}$ for p_i. Hence

$$\boldsymbol{i}_\alpha\boldsymbol{i}_\beta = p_{\alpha\beta,0} + p_{\alpha\beta,1}\boldsymbol{i}_1 + p_{\alpha\beta,2}\boldsymbol{i}_2 + \ldots + p_{\alpha\beta,n}\boldsymbol{i}_n. \tag{5.3}$$

This notation, while somewhat awkward, takes care, at once, of all cases. The choice of the numbers $p_{\alpha\beta,\gamma}$ determines the multiplication table. There are $n^2(n+1)$ such numbers ($n+1$ numbers for each of the n^2 choices of the pairs α, β).

For example, in the case of the complex numbers the multiplication table consists of the single equality

$$\boldsymbol{i} \cdot \boldsymbol{i} = -1 + 0\boldsymbol{i}.$$

In the case of the quaternions, the table contains nine equalities, and can be written down as follows:

	$\boldsymbol{i}$	$\boldsymbol{j}$	$\boldsymbol{k}$
$\boldsymbol{i}$	-1	$\boldsymbol{k}$	$-\boldsymbol{j}$
$\boldsymbol{j}$	$-\boldsymbol{k}$	-1	$\boldsymbol{i}$
$\boldsymbol{k}$	$\boldsymbol{j}$	$-\boldsymbol{i}$	-1

It is clear that each of the nine entries in the table stands for one of the equalities (5.3). For example,

$$\boldsymbol{i}\cdot\boldsymbol{j}=\boldsymbol{k}=0+0\boldsymbol{i}+0\boldsymbol{j}+1\boldsymbol{k}.$$

Given the multiplication table, we define the product

$$(a_0+a_1\boldsymbol{i}_1+\ldots+a_n\boldsymbol{i}_n)(b_0+b_1\boldsymbol{i}_1+\ldots+b_n\boldsymbol{i}_n)$$

to be the result of using the distributive law (each summand in the first sum is multiplied by each summand in the second sum and the results are added). Each term $(a_\alpha\boldsymbol{i}_\alpha)\cdot(b_\beta\boldsymbol{i}_\beta)$ is rewritten as $a_\alpha b_\beta(\boldsymbol{i}_\alpha\boldsymbol{i}_\beta)$, and $\boldsymbol{i}_\alpha\boldsymbol{i}_\beta$ is replaced in accordance with formula (5.3). After reduction we obtain an expression of the form (5.1).

The expressions (5.1) with addition and multiplication defined as above are called a *hypercomplex number system of dimension* $n+1$ and the expressions (5.1) themselves are called *hypercomplex numbers*. It is clear that each hypercomplex number system is completely determined by its multiplication table.

Here are some properties of the multiplication table valid in all hypercomplex number systems.

1. The product of a real number a, viewed as the hypercomplex number $a+0\boldsymbol{i}_1+\ldots+0\boldsymbol{i}_n$, by any number $b_0+b_1\boldsymbol{i}_1+\ldots+b_n\boldsymbol{i}_n$ is obtained by multiplying each of the coefficients $b_0, b_1, \ldots, b_n$ by a:

$$\begin{aligned}(a+0\boldsymbol{i}_1+\ldots+0\boldsymbol{i}_n)&(b_0+b_1\boldsymbol{i}_1+\ldots+b_n\boldsymbol{i}_n)\\&=ab_0+ab_1\boldsymbol{i}_1+\ldots+ab_n\boldsymbol{i}_n,\end{aligned}$$

and

$$\begin{aligned}(b_0+b_1\boldsymbol{i}_1+\ldots+b_n\boldsymbol{i}_n)&(a+0\boldsymbol{i}_1+\ldots+0\boldsymbol{i}_n)\\&=ab_0+ab_1\boldsymbol{i}_1+\ldots+ab_n\boldsymbol{i}_n.\end{aligned}$$

In particular,

$$1\cdot\boldsymbol{v}=\boldsymbol{v} \text{ and } \boldsymbol{v}\cdot 1=\boldsymbol{v},$$

where $\boldsymbol{v}$ is any hypercomplex number.

2. If $\boldsymbol{u}$ and $\boldsymbol{v}$ are hypercomplex numbers then

$$(a\boldsymbol{u})(b\boldsymbol{v})=(ab)(\boldsymbol{u}\boldsymbol{v}),$$

where a and b are any real numbers.

3. The left and right distributive laws hold

$$\boldsymbol{u}(\boldsymbol{v}+\boldsymbol{w}) = \boldsymbol{uv}+\boldsymbol{uw},$$

$$(\boldsymbol{v}+\boldsymbol{w})\boldsymbol{u} = \boldsymbol{vu}+\boldsymbol{wu}.$$

It is clear that properties 1, 2 and 3 are implied by our multiplication rule. We stress that they hold in all hypercomplex number systems.

5.2 Commutative Systems, Associative Systems, and Division Systems

Other "nice" properties of multiplication, such as commutativity

$$\boldsymbol{uv} = \boldsymbol{vu}$$

and associativity

$$(\boldsymbol{uv})\boldsymbol{w} = \boldsymbol{u}(\boldsymbol{vw}),$$

do not necessarily hold in all hypercomplex systems. A system in which for any two elements $\boldsymbol{u}$ and $\boldsymbol{v}$,

$$\boldsymbol{uv} = \boldsymbol{vu}$$

is called a *commutative* system. Of the systems considered so far, the systems of complex, double and dual numbers are commutative and the system of quaternions is not.

▷ It is easy to see the connection between commutativity and the properties of the numbers $p_{\alpha\beta}$ in the multiplication table. If a system is commutative, then

$$\boldsymbol{i}_\alpha \boldsymbol{i}_\beta = \boldsymbol{i}_\beta \boldsymbol{i}_\alpha,$$

that is,

$$p_{\alpha\beta,0} + p_{\alpha\beta,1}\boldsymbol{i}_1 + \ldots + p_{\alpha\beta,n}\boldsymbol{i}_n = p_{\beta\alpha,0} + p_{\beta\alpha,1}\boldsymbol{i}_1 + \ldots + p_{\beta\alpha,n}\boldsymbol{i}_n.$$

Therefore

$$p_{\alpha\beta,0} = p_{\beta\alpha,0},\ \ p_{\alpha\beta,1} = p_{\beta\alpha,1},\ \ \ldots,\ \ p_{\alpha\beta,n} = p_{\beta\alpha,n}, \tag{5.4}$$

where $1 \le \alpha, \beta \le n$. Conversely, if all these equalities hold, then, clearly, the system is commutative. In other words, the system is commutative if and only if the numbers $p_{\alpha\beta,\gamma}$ that determine the multiplication table satisfy the relations (5.4). ◁

A system in which for all triples of numbers $\boldsymbol{u}, \boldsymbol{v}, \boldsymbol{w}$

$$(\boldsymbol{uv})\boldsymbol{w} = \boldsymbol{u}(\boldsymbol{vw}),$$

is called an *associative* system. (The property of associativity is usually part of the definition of a hypercomplex system. In this sense, our definition represents a break with tradition.)

▷ We leave it to the reader to determine the relations among the $p_{\alpha\beta,\gamma}$ that make a hypercomplex system associative. ◁

The systems of complex, double, and dual numbers, as well as the system of quaternions, are associative. A simple example of a nonassociative system is the system of numbers $a+b\boldsymbol{i}+c\boldsymbol{j}$ with the following multiplication table.

$$\boldsymbol{i}^2 = 0, \quad \boldsymbol{j}^2 = 0, \quad \boldsymbol{ji} = 0, \quad \boldsymbol{ij} = \boldsymbol{j}.$$

In this case, $(\boldsymbol{ii})\boldsymbol{j} \neq \boldsymbol{i}(\boldsymbol{ij})$.

Addition, subtraction, and multiplication are defined for all hypercomplex systems. Not so division.

We say that a hypercomplex system is a *division system* (or that it admits division) if for all $\boldsymbol{u}$ and $\boldsymbol{v}$ with $\boldsymbol{v} \neq 0$ the equations

$$\boldsymbol{vx} = \boldsymbol{u}$$

and

$$\boldsymbol{xv} = \boldsymbol{u}$$

are uniquely solvable (for $\boldsymbol{x}$). The solution of the first equation is called the *left quotient* of $\boldsymbol{u}$ by $\boldsymbol{v}$, and the solution of the second equation is called the *right quotient* of $\boldsymbol{u}$ by $\boldsymbol{v}$. In general, the two quotients are different.

The complex numbers and the quaternions are examples of division systems. Their dimensions are 2 and 4, respectively. It is remarkable that the only possible dimensions of a hypercomplex division system are 2,4, and 8. (We shall have more to say about this below.) This suggests that, in the multitude of hypercomplex systems, division systems are few and far between. In particular, the (3-dimensional) system of numbers of the form $a+b\boldsymbol{i}+c\boldsymbol{j}$, with any multiplication table whatever, is not a division system.

Chapter 6

The Doubling Procedure. Cayley Numbers

We shall talk about a remarkable system of hypercomplex numbers called *Cayley numbers.*

Like the complex numbers and the quaternions, the Cayley numbers are a division system, that is, they admit not only addition, subtraction and multiplication, but also division. Also, the Cayley numbers enable us to take a step forward toward the solution of "the sum of squares problem" formulated at the end of chapter 3, in the sense that we obtain the identity (!) for $n = 8$.

Each Cayley number consists of eight terms. It follows that we need seven units $\boldsymbol{i}_1, \boldsymbol{i}_2, \ldots, \boldsymbol{i}_7$ to write down each Cayley number. In other words, the Cayley numbers are expressions of the form

$$a_0 + a_1\boldsymbol{i}_1 + a_2\boldsymbol{i}_2 + a_3\boldsymbol{i}_3 + a_4\boldsymbol{i}_4 + a_5\boldsymbol{i}_5 + a_6\boldsymbol{i}_6 + a_7\boldsymbol{i}_7,$$

where $a_0, a_1, a_2, a_3, a_4, a_5, a_6, a_7$ are arbitrary real numbers.

The rule of multiplication of Cayley numbers is rather involved and we won't state it for a while. Instead, we'll describe a procedure that enables us to construct the Cayley numbers out of the quaternions in a very natural way. We shall call it the *doubling* procedure and say that the Cayley numbers are the result of "doubling" the quaternions. We shall see that the doubling procedure (also called the Cayley-Dickson procedure for the mathematicians Arthur Cayley and Leonard Dickson who first investigated it) can be used not only to obtain the Cayley numbers from the quaternions, but also the quaternions from the complex numbers and the complex numbers from the real numbers.

6.1 Another Approach to the Definition of the Quaternions

Using the fact that $\boldsymbol{ij} = \boldsymbol{k}$ we can write any quaternion

$$\boldsymbol{q} = a + b\boldsymbol{i} + c\boldsymbol{j} + d\boldsymbol{k}$$

in the form

$$\boldsymbol{q} = (a + b\boldsymbol{i}) + (c + d\boldsymbol{i})\boldsymbol{j},$$

or

$$\boldsymbol{q} = \boldsymbol{z}_1 + \boldsymbol{z}_2\boldsymbol{j},$$

where $\boldsymbol{z}_1 = a + b\boldsymbol{i}$, $\boldsymbol{z}_2 = c + d\boldsymbol{i}$.

With this way of writing the quaternions we consider their multiplication.

In addition to $\boldsymbol{q}$ we take a quaternion $\boldsymbol{r}$,

$$\boldsymbol{r} = \boldsymbol{w}_1 + \boldsymbol{w}_2\boldsymbol{j},$$

and consider the product

$$\begin{aligned} \boldsymbol{qr} &= (\boldsymbol{z}_1 + \boldsymbol{z}_2\boldsymbol{j})(\boldsymbol{w}_1 + \boldsymbol{w}_2\boldsymbol{j}) \\ &= \boldsymbol{z}_1\boldsymbol{w}_1 + \boldsymbol{z}_1(\boldsymbol{w}_2\boldsymbol{j}) + (\boldsymbol{z}_2\boldsymbol{j})\boldsymbol{w}_1 + (\boldsymbol{z}_2\boldsymbol{j})(\boldsymbol{w}_2\boldsymbol{j}) \\ &= \boldsymbol{z}_1\boldsymbol{w}_1 + \boldsymbol{z}_1\boldsymbol{w}_2\boldsymbol{j} + \boldsymbol{z}_2\boldsymbol{j}\boldsymbol{w}_1 + \boldsymbol{z}_2\boldsymbol{j}\boldsymbol{w}_2\boldsymbol{j} \qquad (6.1) \end{aligned}$$

(we removed parentheses because quaternion multiplication is associative). We note that since $\boldsymbol{ij} = -\boldsymbol{ji}$, we have $(a + b\boldsymbol{i})\boldsymbol{j} = \boldsymbol{j}(a - b\boldsymbol{i})$, that is,

$$\boldsymbol{zj} = \boldsymbol{j}\bar{\boldsymbol{z}}.$$

Also, it is easy to check that any two elements $\boldsymbol{z}$ and $\boldsymbol{w}$ of the form $a + b\boldsymbol{i}$ commute:

$$\boldsymbol{zw} = \boldsymbol{wz}.$$

With these properties in mind, we can rewrite the second term on the right side of (6.1) as $\boldsymbol{w}_2\boldsymbol{z}_1\boldsymbol{j}$, the third as $\boldsymbol{z}_2\bar{\boldsymbol{w}}_1\boldsymbol{j}$, and the fourth as $\boldsymbol{z}_2\bar{\boldsymbol{w}}_2\boldsymbol{j}^2$ or as $-\bar{\boldsymbol{w}}_2\boldsymbol{z}_2$. It follows that

$$\boldsymbol{qr} = (\boldsymbol{z}_1\boldsymbol{w}_1 - \bar{\boldsymbol{w}}_2\boldsymbol{z}_2) + (\boldsymbol{w}_2\boldsymbol{z}_1 + \boldsymbol{z}_2\bar{\boldsymbol{w}}_1)\boldsymbol{j}. \qquad (6.2)$$

An important point about the representation of a quaternion in the form $\boldsymbol{q} = \boldsymbol{z}_1 + \boldsymbol{z}_2\boldsymbol{j}$ is that, since $\boldsymbol{i}^2 = -1$, all quaternions of the form $a + b\boldsymbol{i}$ may be viewed as complex numbers. This and formula (6.2) justify the following conclusion.

We can define the quaternions as expressions of the form $z_1 + z_2 j$ *where* z_1, *and* z_2 *are complex numbers and* j *is a symbol, that are multiplied as in* (6.2).

This is an essential remark that will enable us to understand the doubling procedure for hypercomplex numbers.

6.2 The Doubling of a Hypercomplex System. Definition of the Cayley Numbers

We introduce a number of definitions. Let $\mathcal{U}$ be a hypercomplex system of elements of the form

$$\boldsymbol{u} = a_0 + a_1\boldsymbol{i}_1 + a_2\boldsymbol{i}_2 + \ldots + a_n\boldsymbol{i}_n$$

with some multiplication rule. We call the element

$$\bar{\boldsymbol{u}} = a_0 - a_1\boldsymbol{i}_1 - a_2\boldsymbol{i}_2 - \ldots - a_n\boldsymbol{i}_n$$

the *conjugate* of $\boldsymbol{u}$.

Now we define $\mathcal{U}^{(2)}$, the *doubled* $\mathcal{U}$, as the hypercomplex system of dimension $2n$ whose elements are expressions of the form

$$\boldsymbol{u}_1 + \boldsymbol{u}_2\boldsymbol{e}, \tag{6.3}$$

where $\boldsymbol{u}_1$ and $\boldsymbol{u}_2$ are arbitrary elements in $\mathcal{U}$ and $\boldsymbol{e}$ is some new symbol. The elements of $\mathcal{U}^{(2)}$ are added according to the natural rule

$$(\boldsymbol{u}_1 + \boldsymbol{u}_2\boldsymbol{e}) + (\boldsymbol{v}_1 + \boldsymbol{v}_2\boldsymbol{e}) = (\boldsymbol{u}_1 + \boldsymbol{v}_1) + (\boldsymbol{u}_2 + \boldsymbol{v}_2)\boldsymbol{e}, \tag{6.4}$$

and multiplied in accordance with the rule

$$(\boldsymbol{u}_1 + \boldsymbol{u}_2\boldsymbol{e})(\boldsymbol{v}_1 + \boldsymbol{v}_2\boldsymbol{e}) = (\boldsymbol{u}_1\boldsymbol{v}_1 - \bar{\boldsymbol{v}}_2\boldsymbol{u}_2) + (\boldsymbol{v}_2\boldsymbol{u}_1 + \boldsymbol{u}_2\bar{\boldsymbol{v}}_1)\boldsymbol{e} \tag{6.5}$$

(the bar denotes conjugation in $\mathcal{U}$).

The reader may be surprised by the fact that in defining the system $\mathcal{U}^{(2)}$ we have ignored the usual method of writing hypercomplex numbers and the use of a multiplication table for the determination of multiplication. As we are about to explain, we have lost nothing and gained brevity and transparency.

The usual form of an element of $\mathcal{U}^{(2)}$ is

$$a_0 + a_1\boldsymbol{i}_1 + \ldots + a_n\boldsymbol{i}_n + a_{n+1}\boldsymbol{i}_{n+1} + \ldots + a_{2n+1}\boldsymbol{i}_{2n+1}. \tag{6.6}$$

This determines the pair of elements $\boldsymbol{u}_1, \boldsymbol{u}_2$ in $\mathcal{U}$ given by

$$\boldsymbol{u}_1 = a_0 + a_1\boldsymbol{i}_1 + \ldots + a_n\boldsymbol{i}_n,$$

$$\boldsymbol{u}_2 = a_{n+1} + a_{n+2}\boldsymbol{i}_1 + \ldots + a_{2n+1}\boldsymbol{i}_n,$$

and thus the element(6.3) (which may be regarded as a brief code for (6.6)), and conversely. Beyond that, the definition of multiplication (6.5) is shorter and clearer than a definition in terms of a multiplication table. Of course, formula (6.5) can be used to obtain the multiplication table of the "imaginary units" $\boldsymbol{i}_1, \boldsymbol{i}_2, \ldots, \boldsymbol{i}_{2n+1}$. We won't produce such a table in general but will give it in detail below for the Cayley numbers.

Now that we have defined the doubling procedure it is easy to see that what we did in the beginning of this chapter is obtain the quaternions by doubling the complex numbers. We leave it to the reader to show that doubling the real numbers yields the complex numbers.

As stated earlier, the main purpose of this chapter is to construct the system of Cayley numbers. *We can now define the Cayley numbers as the system obtained by doubling the quaternions.* All the properties of the Cayley numbers flow naturally from this definition. They will be studied in detail in the next section.

6.3 The Multiplication Table of the Cayley Numbers

By definition, Cayley numbers are numbers of the form

$$\boldsymbol{q}_1 + \boldsymbol{q}_2\boldsymbol{e},$$

where $\boldsymbol{q}_1$ and $\boldsymbol{q}_2$ are arbitrary quaternions, that are multiplied in accordance with the rule

$$(\boldsymbol{q}_1 + \boldsymbol{q}_2\boldsymbol{e})(\boldsymbol{r}_1 + \boldsymbol{r}_2\boldsymbol{e}) = (\boldsymbol{q}_1\boldsymbol{r}_1 - \bar{\boldsymbol{r}}_2\boldsymbol{q}_2) + (\boldsymbol{r}_2\boldsymbol{q}_1 + \boldsymbol{q}_2\bar{\boldsymbol{r}}_1)\boldsymbol{e}. \tag{6.7}$$

We consider the connection between this definition of the Cayley numbers and their representation in the form

$$a_0 + a_1\boldsymbol{i}_1 + a_2\boldsymbol{i}_2 + a_3\boldsymbol{i}_3 + a_4\boldsymbol{i}_4 + a_5\boldsymbol{i}_5 + a_6\boldsymbol{i}_6 + a_7\boldsymbol{i}_7. \tag{6.8}$$

More precisely, we construct the multiplication table of the "imaginary units" $\boldsymbol{i}_1, \ldots, \boldsymbol{i}_7$.

The quaternions $\boldsymbol{q}_1$ and $\boldsymbol{q}_2$ corresponding to the representation (6.8) are

$$\boldsymbol{q}_1 = a_0 + a_1\boldsymbol{i} + a_2\boldsymbol{j} + a_3\boldsymbol{k}, \ \boldsymbol{q}_2 = a_4 + a_5\boldsymbol{i} + a_6\boldsymbol{j} + a_7\boldsymbol{k}.$$

For greater uniformity, we shall write (6.8) as

$$a + bi + cj + dk + AE + BI + CJ + DK,$$

where a, b, c, d, A, B, C, D are the earlier letters $a_0, a_1, \ldots, a_7$ and $\boldsymbol{i}, \boldsymbol{j}, \boldsymbol{k}, \boldsymbol{E}, \boldsymbol{I}, \boldsymbol{J}, \boldsymbol{K}$ are new symbols for the imaginary units $\boldsymbol{i}_1, \boldsymbol{i}_2, \ldots, \boldsymbol{i}_7$.

Now the quaternions $\boldsymbol{q}_1$ and $\boldsymbol{q}_2$ are written as

$$\boldsymbol{q}_1 = a + b\boldsymbol{i} + c\boldsymbol{j} + d\boldsymbol{k}, \ \boldsymbol{q}_2 = A + B\boldsymbol{i} + C\boldsymbol{j} + D\boldsymbol{k}.$$

Beginning with(6.7) we can, as noted earlier, construct the multiplication table for the units $\boldsymbol{i}, \boldsymbol{j}, \boldsymbol{k}, \boldsymbol{E}, \boldsymbol{I}, \boldsymbol{J}, \boldsymbol{K}$. For example, if in (6.7) we put $\boldsymbol{q}_2 = \boldsymbol{r}_2 = 0$, then

$$(\boldsymbol{q}_1 + 0\boldsymbol{e})(\boldsymbol{r}_1 + 0\boldsymbol{e}) = \boldsymbol{q}_1\boldsymbol{r}_1 + 0\boldsymbol{e}.$$

Thus the Cayley numbers $\boldsymbol{q}_1$ and $\boldsymbol{r}_1$ multiply like quaternions. It follows that the multiplication table for the units $\boldsymbol{i}, \boldsymbol{j}, \boldsymbol{k}$ is the same as for the quaternions:

$$\begin{aligned} \boldsymbol{i}^2 = -1, \ \boldsymbol{j}^2 &= -1, \ \boldsymbol{k}^2 = -1, \\ \boldsymbol{ij} = \boldsymbol{k}, &\ \ \boldsymbol{ji} = -\boldsymbol{k}, \\ \boldsymbol{jk} = \boldsymbol{i}, &\ \ \boldsymbol{kj} = -\boldsymbol{i}, \\ \boldsymbol{ki} = \boldsymbol{j}, &\ \ \boldsymbol{ik} = -\boldsymbol{j}. \end{aligned}$$

This gives 9 of the 49 products (there are $7 \cdot 7 = 49$ products of the 7 units). Instead of a table of the remaining 40 products we give a mnemonic scheme for remembering the whole table. First there is the collection of seven triples:

	$\boldsymbol{I}$	$\boldsymbol{J}$	$-\boldsymbol{k}$	$\boldsymbol{i}$	$\boldsymbol{E}$	$\boldsymbol{I}$
$\boldsymbol{i}\ \boldsymbol{j}\ \boldsymbol{k}$	$\boldsymbol{I}$	$-\boldsymbol{j}$	$\boldsymbol{K}$	$\boldsymbol{j}$	$\boldsymbol{E}$	$\boldsymbol{J}$
	$-\boldsymbol{i}$	$\boldsymbol{J}$	$\boldsymbol{K}$	$\boldsymbol{k}$	$\boldsymbol{E}$	$\boldsymbol{K}$

To remember them, note that the triples in the left frame are obtained from the triple $\boldsymbol{i}, \boldsymbol{j}, \boldsymbol{k}$ by putting a minus sign in front of one symbol and

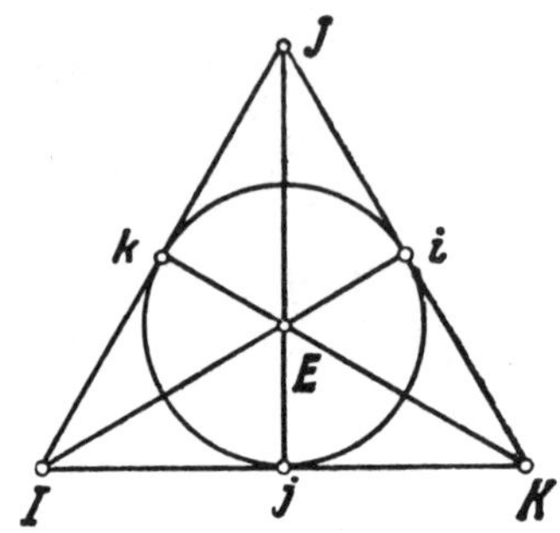

Figure 6.1.

capitalizing the other two. In all triples in the right frame the middle symbol is an $\boldsymbol{E}$ and the other two symbols are the same except for the type. To multiply them, let α, β, γ denote any one of the 7 triples. Put

$$\begin{aligned} \alpha\beta &= \gamma, & \beta\alpha &= -\gamma, \\ \beta\gamma &= \alpha, & \gamma\beta &= -\alpha, \\ \gamma\alpha &= \beta, & \alpha\gamma &= -\beta, \end{aligned}$$

and

$$a^2 = -1,\ \beta^2 = -1,\ \gamma^2 = -1,$$

that is, α, β, γ multiply like the quaternions $\boldsymbol{i}, \boldsymbol{j}, \boldsymbol{k}$.

▷ Figure 6.1 provides a good illustration of our rule. It shows a triangle with vertices $\boldsymbol{I}, \boldsymbol{J}, \boldsymbol{K}$ whose medians meet the sides at $\boldsymbol{i}, \boldsymbol{j}, \boldsymbol{k}$ and intersect at $\boldsymbol{E}$. There are three "imaginary" units on each line. The units $\boldsymbol{i}, \boldsymbol{j}, \boldsymbol{k}$ also lie on a "line" (represented by the circle). In all, there are 7 "lines" and three units on each line. Apart from sign, the product of two units is the unit "collinear" with them.

It is of interest to point out that in order to obtain a correct multiplication scheme for the units $\boldsymbol{i}, \boldsymbol{j}, \boldsymbol{k}, \boldsymbol{E}, \boldsymbol{I}, \boldsymbol{J}, \boldsymbol{K}$ it suffices to place (in this figure) $\boldsymbol{i}, \boldsymbol{j}, \boldsymbol{k}$ on any line, mark one of the remaining points $\boldsymbol{E}$, and place $\boldsymbol{I}, \boldsymbol{J}, \boldsymbol{K}$ on the lines $\boldsymbol{iE}, \boldsymbol{jE}, \boldsymbol{kE}$, respectively. ◁

6.4 Conjugation of Cayley Numbers. Absolute Values of Cayley Numbers

Let

$$u = a + bi + cj + dk + AE + BI + CJ + DK \tag{6.9}$$

be any Cayley number. By its *conjugate* we mean the Cayley number

$$\bar{u} = a - bi - cj - dk - AE - BI - CJ - DK.$$

If instead of (6.9) we use the short representation

$$u = q_1 + q_2 e,$$

where

$$q_1 = a + bi + cj + dk, \quad q_2 = A + Bi + Cj + Dk,$$

then the conjugate Cayley number is given by

$$\bar{u} = \bar{q}_1 - q_2 e.$$

Now we compute the product of any Cayley number u and its conjugate $\bar{u}$. It turns out that, just as in the case of complex numbers and quaternions, this product is a real number (that is, a Cayley number of the form $a + 0i + 0j + \ldots + 0K$). In fact,

$$u\bar{u} = (q_1 + q_2 e)(\bar{q}_1 - q_2 e) = (q_1\bar{q}_1 + \bar{q}_2 q_2) + (-q_2 q_1 + q_2 q_1)e.$$

Bearing in mind that for quaternions $q\bar{q} = \bar{q}q = |q|^2$, we see that

$$u\bar{u} = q_1\bar{q}_1 + q_2\bar{q}_2 = |q_1|^2 + |q_2|^2. \tag{6.10}$$

The square root of $|q_1|^2 + |q_2|^2$ is called the *absolute value* or *norm* of the Cayley number u and is denoted by $|u|$. Note that if u is given in the form (6.9), then the square of its absolute value is

$$a^2 + b^2 + c^2 + d^2 + A^2 + B^2 + C^2 + D^2. \tag{6.11}$$

In view of the definition of the absolute value we have

$$u\bar{u} = |u|^2. \tag{6.12}$$

If we bear in mind that the squares of the absolute values of the Cayley numbers u and $\bar{u}$ are equal (in fact, both are equal to (6.11)), then we also have

$$\bar{u}u = |u|^2.$$

6.5 The Absolute Value of the Product of Cayley Numbers

The system of Cayley numbers shares many of the properties of the systems of complex numbers and quaternions. One such property is that the absolute value of the product of Cayley numbers is the product of the absolute values of the factors:

$$|\boldsymbol{u}\boldsymbol{v}| = |\boldsymbol{u}||\boldsymbol{v}|, \tag{6.13}$$

or, equivalently,

$$|\boldsymbol{u}\boldsymbol{v}|^2 = |\boldsymbol{u}|^2|\boldsymbol{v}|^2. \tag{6.14}$$

We prove (6.14) by direct computation of $|\boldsymbol{u}\boldsymbol{v}|^2$ and $|\boldsymbol{u}|^2|\boldsymbol{v}|^2$. If we apply formula (6.10) to the product

$$\boldsymbol{u}\boldsymbol{v} = (\boldsymbol{q}_1 + \boldsymbol{q}_2\boldsymbol{e})(\boldsymbol{r}_1 + \boldsymbol{r}_2\boldsymbol{e}) = (\boldsymbol{q}_1\boldsymbol{r}_1 - \bar{\boldsymbol{r}}_2\boldsymbol{q}_2) + (\boldsymbol{r}_2\boldsymbol{q}_1 + \boldsymbol{q}_2\bar{\boldsymbol{r}}_1)\boldsymbol{e},$$

then we obtain

$$|\boldsymbol{u}\boldsymbol{v}|^2 = (\boldsymbol{q}_1\boldsymbol{r}_1 - \bar{\boldsymbol{r}}_2\boldsymbol{q}_2)(\overline{\boldsymbol{q}_1\boldsymbol{r}_1 - \bar{\boldsymbol{r}}_2\boldsymbol{q}_2}) + (\boldsymbol{r}_2\boldsymbol{q}_1 + \boldsymbol{q}_2\bar{\boldsymbol{r}}_1)(\overline{\boldsymbol{r}_2\boldsymbol{q}_1 + \boldsymbol{q}_2\bar{\boldsymbol{r}}_1}),$$

or, in view of the property of conjugation for quaternions,

$$|\boldsymbol{u}\boldsymbol{v}|^2 = (\boldsymbol{q}_1\boldsymbol{r}_1 - \bar{\boldsymbol{r}}_2\boldsymbol{q}_2)(\bar{\boldsymbol{r}}_1\bar{\boldsymbol{q}}_1 - \bar{\boldsymbol{q}}_2\boldsymbol{r}_2) + (\boldsymbol{r}_2\boldsymbol{q}_1 + \boldsymbol{q}_2\bar{\boldsymbol{r}}_1)(\bar{\boldsymbol{q}}_1\bar{\boldsymbol{r}}_2 + \boldsymbol{r}_1\bar{\boldsymbol{q}}_2).$$

On the other hand,

$$|\boldsymbol{u}|^2|\boldsymbol{v}|^2 = (\boldsymbol{q}_1\bar{\boldsymbol{q}}_1 + \boldsymbol{q}_2\bar{\boldsymbol{q}}_2)(\boldsymbol{r}_1\bar{\boldsymbol{r}}_1 + \boldsymbol{r}_2\bar{\boldsymbol{r}}_2).$$

If we compare the two expressions, then we see that they differ by the sum S of four terms,

$$S = \boldsymbol{r}_2\boldsymbol{q}_1\boldsymbol{r}_1\bar{\boldsymbol{q}}_2 + \boldsymbol{q}_2\bar{\boldsymbol{r}}_1\bar{\boldsymbol{q}}_1\bar{\boldsymbol{r}}_2 - \boldsymbol{q}_1\boldsymbol{r}_1\bar{\boldsymbol{q}}_2\boldsymbol{r}_2 - \bar{\boldsymbol{r}}_2\boldsymbol{q}_2\bar{\boldsymbol{r}}_1\bar{\boldsymbol{q}}_1.$$

Therefore we must show that $S = 0$ for any four quaternions $\boldsymbol{q}_1, \boldsymbol{q}_2, \boldsymbol{r}_1, \boldsymbol{r}_2$.

We begin with the obvious observation that $S = 0$ if $\boldsymbol{r}_2$ is real. On the other hand, if $\boldsymbol{r}_2$ is a pure imaginary quaternion, then $\bar{\boldsymbol{r}}^2 = -\boldsymbol{r}_2$ and

$$S = \boldsymbol{r}_2(\boldsymbol{q}_1\boldsymbol{r}_1\bar{\boldsymbol{q}}_2 + \boldsymbol{q}_2\bar{\boldsymbol{r}}_1\bar{\boldsymbol{q}}_1) - (\boldsymbol{q}_1\boldsymbol{r}_1\bar{\boldsymbol{q}}_2 + \boldsymbol{q}_2\bar{\boldsymbol{r}}_1\bar{\boldsymbol{q}}_1)\boldsymbol{r}_2.$$

The expression in parentheses is a sum of two conjugate quaternions and therefore equal to some real number c. Hence

$$S = \boldsymbol{r}_2 c - c\boldsymbol{r}_2 = 0.$$

It remains to note an obvious property of S: if it vanishes for $\boldsymbol{r}_2 = \boldsymbol{a}$ and $\boldsymbol{r}_2 = \boldsymbol{b}$, then it also vanishes for $\boldsymbol{r}_2 = \boldsymbol{a} + \boldsymbol{b}$. Since every quaternion is a sum of a real number and a pure imaginary quaternion and for each of these $S = 0$, it follows that S is always equal to zero.

6.6 The Eight-Square Identity

The identity

$$|\boldsymbol{uv}|^2 = |\boldsymbol{u}|^2|\boldsymbol{v}|^2 \tag{6.15}$$

established in the previous section presents a new contribution to the solution of "the problem of the sum of squares" posed at the end of chapter 3. Indeed, if we write it out in detail (and read it from right to left), then this identity states that *"the product of sums of eight squares is again a sum of eight squares."* In fact, if

$$\boldsymbol{u} = a + b\boldsymbol{i} + c\boldsymbol{j} + d\boldsymbol{k} + A\boldsymbol{E} + B\boldsymbol{I} + C\boldsymbol{J} + D\boldsymbol{K},$$

$$\boldsymbol{v} = a' + b'\boldsymbol{i} + c'\boldsymbol{j} + d'\boldsymbol{k} + A'\boldsymbol{E} + B'\boldsymbol{I} + C'\boldsymbol{J} + D'\boldsymbol{K},$$

and

$$\boldsymbol{uv} = \Phi_0 + \Phi_1\boldsymbol{i} + \Phi_2\boldsymbol{j} + \Phi_3\boldsymbol{k} + \Phi_4\boldsymbol{E} + \Phi_5\boldsymbol{I} + \Phi_6\boldsymbol{J} + \Phi_7\boldsymbol{K},$$

then the identity (6.15) takes the form

$$(a^2 + \ldots + D^2)(a'^2 + \ldots + D'^2) = \Phi_0^2 + \Phi_1^2 + \ldots + \Phi_7^2.$$

Of course, we must make use of the multiplication of Cayley numbers to express $\Phi_0, \Phi_1, \ldots, \Phi_7$ in terms of $a, \ldots, D$, $a', \ldots, D'$. This tedious task yields the identity:

$$\begin{aligned}
&(a^2 + b^2 + c^2 + d^2 + A^2 + B^2 + C^2 + D^2)\\
&\quad \times (a'^2 + b'^2 + c'^2 + d'^2 + A'^2 + B'^2 + C'^2 + D'^2)\\
&\quad = (aa' - bb' - cc' - dd' - AA' - BB' - CC' - DD')^2\\
&\quad +(ab' + ba' + cd' - dc' - A'B + B'A + C'D - D'C)^2\\
&\quad +(ac' + ca' - bd' + db' - A'C + C'A - B'D + D'B)^2\\
&\quad +(ad' + da' + bc' - cb' - A'D + D'A + B'C - C'B)^2\\
&\quad +(A'a - B'b - C'c - D'd + Aa' + Bb' + Cc' + Dd')^2\\
&\quad +(A'b + B'a + C'd - D'c - Ab' + Ba' - Cd' + Dc')^2\\
&\quad +(A'c + C'a - B'd + D'b - Ac' + Ca' + Bd' - Db')^2\\
&\quad +(A'd + D'a + B'c - C'b - Ad' + Da' - Bc' + Cb')^2.
\end{aligned}$$

It is of interest to note that it was the search for an eight-square identity that led the English mathematician Cayley to the discovery of the Cayley numbers!

6.7 The Non-associativity of Cayley Numbers. The Alternative Property

We said earlier that the Cayley numbers share many, but not all, of the properties of quaternions and complex numbers. Thus, whereas multiplication of complex numbers and quaternions is associative, *multiplication of Cayley numbers is not associative.* For example,

$$(\boldsymbol{i}\boldsymbol{j})\boldsymbol{E} \neq \boldsymbol{i}(\boldsymbol{j}\boldsymbol{E}).$$

Indeed, $(\boldsymbol{i}\boldsymbol{j})\boldsymbol{E} = \boldsymbol{k}\boldsymbol{E} = \boldsymbol{K}$, and $\boldsymbol{i}(\boldsymbol{j}\boldsymbol{E}) = \boldsymbol{i}\boldsymbol{J} = -\boldsymbol{K}$.

Obviously, the nonassociativity of multiplication of Cayley numbers does not mean that for any three such numbers $\boldsymbol{u}, \boldsymbol{v}, \boldsymbol{w}$ we have $(\boldsymbol{u}\boldsymbol{v})\boldsymbol{w} \neq \boldsymbol{u}(\boldsymbol{v}\boldsymbol{w})$. In fact, we will show that the following equalities hold for any two Cayley numbers $\boldsymbol{u}, \boldsymbol{v}$:

$$(\boldsymbol{u}\boldsymbol{v})\boldsymbol{v} = \boldsymbol{u}(\boldsymbol{v}\boldsymbol{v}), \tag{6.16}$$

and

$$\boldsymbol{v}(\boldsymbol{v}\boldsymbol{u}) = (\boldsymbol{v}\boldsymbol{v})\boldsymbol{u}. \tag{6.17}$$

We can regard formulas (6.16) and (6.17) as a weak form of associativity. Systems in which these two formulas hold are called *alternative* systems.

Note that instead of proving (6.16) and (6.17) it suffices to prove

$$(\boldsymbol{u}\boldsymbol{v})\bar{\boldsymbol{v}} = \boldsymbol{u}(\boldsymbol{v}\bar{\boldsymbol{v}}), \tag{6.16$'$}$$

and

$$\bar{\boldsymbol{v}}(\boldsymbol{v}\boldsymbol{u}) = (\bar{\boldsymbol{v}}\boldsymbol{v})\boldsymbol{u}. \tag{6.17$'$}$$

Indeed, if we replace $\bar{\boldsymbol{v}}$ in these equalities by $-\boldsymbol{v} + 2a$, where a is the real part of the Cayley number $\boldsymbol{v}$, then we can easily obtain (6.16) and (6.17).

We prove (6.16′). A similar proof establishes (6.17′).

Put $\boldsymbol{u} = \boldsymbol{q}_1 + \boldsymbol{q}_2\boldsymbol{e}, \quad \boldsymbol{v} = \boldsymbol{r}_1 + \boldsymbol{r}_2\boldsymbol{e}$. Then

$$\begin{aligned}
(\boldsymbol{u}\boldsymbol{v})\bar{\boldsymbol{v}} &= ((\boldsymbol{q}_1 + \boldsymbol{q}_2\boldsymbol{e})(\boldsymbol{r}_1 + \boldsymbol{r}_2\boldsymbol{e}))(\bar{\boldsymbol{r}}_1 - \boldsymbol{r}_2\boldsymbol{e}) \\
&= ((\boldsymbol{q}_1\boldsymbol{r}_1 - \bar{\boldsymbol{r}}_2\boldsymbol{q}_2) + (\boldsymbol{r}_2\boldsymbol{q}_1 + \boldsymbol{q}_2\bar{\boldsymbol{r}}_1)\boldsymbol{e})(\bar{\boldsymbol{r}}_1 - \boldsymbol{r}_2\boldsymbol{e}) \\
&= ((\boldsymbol{q}_1\boldsymbol{r}_1 - \bar{\boldsymbol{r}}_2\boldsymbol{q}_2)\bar{\boldsymbol{r}}_1 + \bar{\boldsymbol{r}}_2(\boldsymbol{r}_2\boldsymbol{q}_1 + \boldsymbol{q}_2\bar{\boldsymbol{r}}_1)) \\
&+ ((-\boldsymbol{r}_2)(\boldsymbol{q}_1\boldsymbol{r}_1 - \bar{\boldsymbol{r}}_2\boldsymbol{q}_2) + (\boldsymbol{r}_2\boldsymbol{q}_1 + \boldsymbol{q}_2\bar{\boldsymbol{r}}_1)\boldsymbol{r}_1)\boldsymbol{e} \\
&= (|\boldsymbol{r}_1|^2 + |\boldsymbol{r}_2|^2)\boldsymbol{q}_1 + (|\boldsymbol{r}_1|^2 + |\boldsymbol{r}_2|^2)\boldsymbol{q}_2\boldsymbol{e} \\
&= (|\boldsymbol{r}_1|^2 + |\boldsymbol{r}_2|^2)(\boldsymbol{q}_1 + \boldsymbol{q}_2\boldsymbol{e}) = |\boldsymbol{v}|^2\boldsymbol{u}.
\end{aligned}$$

On the other hand, $\boldsymbol{v}\bar{\boldsymbol{v}} = |\boldsymbol{v}|^2$, so that

$$\boldsymbol{u}(\boldsymbol{v}\bar{\boldsymbol{v}}) = |\boldsymbol{v}|^2\boldsymbol{u}.$$

This implies (6.16′).

6.8 The Cayley Numbers Are a Division System

Like the complex numbers and the quaternions, the Cayley numbers are a *division system*. Let $\boldsymbol{u}$ and $\boldsymbol{v}$ be any two Cayley numbers and $\boldsymbol{v} \neq 0$. We recall that the left quotient of $\boldsymbol{u}$ by $\boldsymbol{v}$ is the solution of the equation

$$\boldsymbol{v}\boldsymbol{x} = \boldsymbol{u}, \tag{6.18}$$

and the right quotient of $\boldsymbol{u}$ by $\boldsymbol{v}$ is the solution of the equation

$$\boldsymbol{x}\boldsymbol{v} = \boldsymbol{u}. \tag{6.19}$$

We solve (6.18). Just as in the case of the quaternions, we multiply both sides of (6.18) on the left by $\bar{\boldsymbol{v}}$. This yields

$$\bar{\boldsymbol{v}}(\boldsymbol{v}\boldsymbol{x}) = \bar{\boldsymbol{v}}\boldsymbol{u},$$

or, in view of (6.17′),

$$|\boldsymbol{v}|^2\boldsymbol{x} = \boldsymbol{v}\boldsymbol{u}.$$

Hence

$$\boldsymbol{x} = \bar{\boldsymbol{v}}\boldsymbol{u}/|\boldsymbol{v}|^2.$$

Direct substitution (and the use of (6.17′)) shows that this value of $\boldsymbol{x}$ satisfies (6.18). In other words, the left quotient of $\boldsymbol{u}$ by $\boldsymbol{v}$ is

$$\boldsymbol{x}_l = \bar{\boldsymbol{v}}\boldsymbol{u}/|\boldsymbol{v}|^2.$$

A similar argument shows that the right quotient is

$$\boldsymbol{x}_r = \boldsymbol{u}\bar{\boldsymbol{v}}/|\boldsymbol{v}|^2.$$

(the proof requires the use of formula (6.16′)).

Thus we have shown that the Cayley numbers are indeed a *division system*.

Chapter 7

Algebras

7.1 Heuristic Considerations

Let us go back to the concept of a hypercomplex system. According to the definition in chapter 5, a hypercomplex system of dimension $n+1$ is the set of expressions

$$a_0 + a_1\boldsymbol{i}_1 + a_2\boldsymbol{i}_2 + \ldots + a_n\boldsymbol{i}_n$$

(hypercomplex numbers) with a natural rule of addition and a certain rule of multiplication. The latter is determined by prescribing a table of products

$$\boldsymbol{i}_\alpha\boldsymbol{i}_\beta = p_{\alpha\beta,0} + p_{\alpha\beta,1}\boldsymbol{i}_1 + \ldots + p_{\alpha\beta,n}\boldsymbol{i}_n \tag{7.1}$$

of the "imaginary units" $\boldsymbol{i}_1, \boldsymbol{i}_2, \ldots, \boldsymbol{i}_n$, and stipulating that the product of two hypercomplex numbers is obtained by using the distributive laws, and by replacing $(a_\alpha\boldsymbol{i}_\alpha)(b_\beta\boldsymbol{i}_\beta)$ by $a_\alpha b_\beta(\boldsymbol{i}_\alpha\boldsymbol{i}_\beta)$ and $\boldsymbol{i}_\alpha\boldsymbol{i}_\beta$ by the right side of (7.1).

We consider the case when all numbers $p_{\alpha\beta,0}$ (the "free terms" in formula (7.1)) are zero. In that case the product of two imaginary units $\boldsymbol{i}_\alpha, \boldsymbol{i}_\beta$ is again a combination of imaginary units.

Let $\mathcal{A}$ denote the set of hypercomplex numbers of the form

$$\boldsymbol{a} = a_1\boldsymbol{i}_1 + a_2\boldsymbol{i}_2 + \ldots + a_n\boldsymbol{i}_n \tag{7.2}$$

(without the free terms). It is clear that the sum of two such numbers is again a number of the form (7.2). By what has been said about the products $\boldsymbol{i}_\alpha\boldsymbol{i}_\beta$, it follows that the product of two numbers of the form (7.2) is also a number of that form. Thus the set $\mathcal{A}$ is closed under addition and multiplication. This allows us to consider $\mathcal{A}$ as an *independent system* with two operations, addition and multiplication. In general, $\mathcal{A}$ is not a hypercomplex system in the sense of our use of the word (the case when $\mathcal{A}$ can be regarded as such will be discussed below).

The main difference between the system $\mathcal{A}$ and a hypercomplex system is that the latter contains an element $\boldsymbol{e}$ such that

$$\boldsymbol{e} \cdot \boldsymbol{a} = \boldsymbol{a} \cdot \boldsymbol{e} = \boldsymbol{a}$$

for all $\boldsymbol{a}$ (the $\boldsymbol{e}$ in question is the element $1+0\boldsymbol{i}_1+\ldots+0\boldsymbol{i}_n$) and the former, in general, does not. A closely related difference is that in a hypercomplex system it makes sense to speak of the product of a real number k by any element $\boldsymbol{a}$ (by definition, this is the product of $k = k + 0\boldsymbol{i}_1 + \ldots + 0\boldsymbol{i}_n$ and $\boldsymbol{a}$) whereas in $\mathcal{A}$ the symbol $k\boldsymbol{a}$ is meaningless. The latter difference can easily be set aside by simply *defining* the product of a real number by an element of $\mathcal{A}$ by means of the formula

$$k(a_1\boldsymbol{i}_1 + a_2\boldsymbol{i}_2 + \ldots + a_n\boldsymbol{i}_n) = ka_1\boldsymbol{i}_1 + ka_2\boldsymbol{i}_2 + \ldots + ka_n\boldsymbol{i}_n.$$

In this way, the set $\mathcal{A}$ (on which one has already defined addition and multiplication) becomes an object called an *n-dimensional algebra*, or simply an *algebra* (not to be confused with the branch of mathematics bearing the same name!).

7.2 Definition of an Algebra

By an *n-dimensional algebra* we mean the set of expressions of the form

$$a_1\boldsymbol{i}_1 + a_2\boldsymbol{i}_2 + \ldots + a_n\boldsymbol{i}_n \tag{7.2}$$

(where $a_1, a_2, \ldots, a_n$ are arbitrary real numbers and $\boldsymbol{i}_1, \boldsymbol{i}_2, \ldots, \boldsymbol{i}_n$ are certain symbols) with the following operations:

1. multiplication by a real number

$$k(a_1\boldsymbol{i}_1 + a_2\boldsymbol{i}_2 + \ldots + a_n\boldsymbol{i}_n) = ka_1\boldsymbol{i}_1 + ka_2\boldsymbol{i}_2 + \ldots + ka_n\boldsymbol{i}_n; \tag{7.3}$$

2. addition

$$(a_1\boldsymbol{i}_1 + a_2\boldsymbol{i}_2 + \ldots + a_n\boldsymbol{i}_n) + (b_1\boldsymbol{i}_1 + b_2\boldsymbol{i}_2 + \ldots + b_n\boldsymbol{i}_n) = \\ = (a_1 + b_1)\boldsymbol{i}_1 + (a_2 + b_2)\boldsymbol{i}_2 + \ldots + (a_n + b_n)\boldsymbol{i}_n; \tag{7.4}$$

3. multiplication given in terms of *a* table of products

$$\boldsymbol{i}_\alpha\boldsymbol{i}_\beta = p_{\alpha\beta,1}\boldsymbol{i}_1 + p_{\alpha\beta,2}\boldsymbol{i}_2 + \ldots + p_{\alpha\beta,n}\boldsymbol{i}_n, \tag{7.5}$$

where α and β are integers from 1 to n (the table is used to find the product

$$(a_1\boldsymbol{i}_1 + a_2\boldsymbol{i}_2 + \ldots + a_n\boldsymbol{i}_n)(b_1\boldsymbol{i}_1 + b_2\boldsymbol{i}_2 + \ldots + b_n\boldsymbol{i}_n)$$

just as in the case of hypercomplex numbers).

Our definition makes it clear that an n-dimensional algebra is completely determined by its "multiplication table"(7.5), that is, by the choice of n^3 numbers $p_{\alpha\beta,\gamma}$. In principle, these numbers are not subject to any restrictions; each choice determines a certain algebra.

7.3 A Hypercomplex System as a Special Case of an Algebra

For the sake of clarity, we "extracted" the concept of an algebra from that of a hypercomplex system. This fact notwithstanding, it must be emphasized that the concept of an algebra is more general than that of a hypercomplex system in the sense that *every hypercomplex system can be regarded as an algebra of the same dimension.* A detailed explanation follows.

Let $\mathcal{A}$ be an algebra with elements

$$a_1\boldsymbol{i}_1 + a_2\boldsymbol{i}_2 + \ldots + a_n\boldsymbol{i}_n$$

and multiplication table

$$\boldsymbol{i}_\alpha\boldsymbol{i}_\beta = p_{\alpha\beta,1}\boldsymbol{i}_1 + p_{\alpha\beta,2}\boldsymbol{i}_2 + \ldots + p_{\alpha\beta,n}\boldsymbol{i}_n$$

(α, β are numbers from 1 to n), in which the unit $\boldsymbol{i}_1$ has the property

$$\boldsymbol{i}_1\boldsymbol{i}_\alpha = \boldsymbol{i}_\alpha \text{ and } \boldsymbol{i}_\alpha\boldsymbol{i}_1 = \boldsymbol{i}_\alpha \tag{7.6}$$

for all α from 1 to n. Together with this algebra we consider the hypercomplex system of elements

$$a_1 + a_2\boldsymbol{i}_2 + \ldots + a_n\boldsymbol{i}_n$$

with the multiplication table

$$\boldsymbol{i}_\alpha\boldsymbol{i}_\beta = p_{\alpha\beta,1} + p_{\alpha\beta,2}\boldsymbol{i}_2 + \ldots + p_{\alpha\beta,n}\boldsymbol{i}_n$$

(α, β are numbers from 1 to n). We shall say that this hypercomplex system *corresponds* to the algebra $\mathcal{A}$.

Using the multiplication table, we can find the product of any two elements of our algebra:

$$(a_1\boldsymbol{i}_1 + a_2\boldsymbol{i}_2 + \ldots + a_n\boldsymbol{i}_n)(b_1\boldsymbol{i}_1 + b_2\boldsymbol{i}_2 + \ldots + b_n\boldsymbol{i}_n)$$
$$= c_1\boldsymbol{i}_1 + c_2\boldsymbol{i}_2 + \ldots + c_n\boldsymbol{i}_n.$$

If we "clear away" $\boldsymbol{i}_1$ in the last equality, then we obtain the relation

$$(a_1 + a_2\boldsymbol{i}_2 + \ldots + a_n\boldsymbol{i}_n)(b_1 + b_2\boldsymbol{i}_2 + \ldots + b_n\boldsymbol{i}_n)$$
$$= c_1 + c_2\boldsymbol{i}_2 + \ldots + c_n\boldsymbol{i}_n,$$

which coincides with the law of multiplication in the corresponding hypercomplex system. It follows that by starting with an algebra satisfying condition (7.6) and "clearing away" the symbol $\boldsymbol{i}_1$ in the representation of its elements, we obtain a hypercomplex system of the same dimension. Moreover, since the numbers $p_{\alpha\beta,\gamma}$ for $\alpha > 1, \beta > 1$ are arbitrary, we can obtain in this way *all* hypercomplex system. For example, consider the two-dimensional algebra with the multiplication table

$$\boldsymbol{i}_1\boldsymbol{i}_1 = \boldsymbol{i}_1, \quad \boldsymbol{i}_1\boldsymbol{i}_2 = \boldsymbol{i}_2, \quad \boldsymbol{i}_2\boldsymbol{i}_1 = \boldsymbol{i}_2, \quad \boldsymbol{i}_2\boldsymbol{i}_2 = -\boldsymbol{i}_1.$$

Clearly, this algebra satisfies condition (7.6). If in the product

$$(a_1\boldsymbol{i}_1 + a_2\boldsymbol{i}_2)(b_1\boldsymbol{i}_1 + b_2\boldsymbol{i}_2) = (a_1b_1 - a_2b_2)\boldsymbol{i}_1 + (a_1b_2 + a_2b_1)\boldsymbol{i}_2$$

of two elements of $\mathcal{A}$ we strike out $\boldsymbol{i}_1$, then the multiplication table reduces to

$$\boldsymbol{i}_2\boldsymbol{i}_2 = -1,$$

and the multiplication rule becomes

$$(a_1 + a_2\boldsymbol{i}_2)(b_1 + b_2\boldsymbol{i}_2) = (a_1b_1 - a_2b_2) + (a_1b_2 + a_2b_1)\boldsymbol{i}_2,$$

which shows that we are dealing, essentially, with the system of complex numbers.

7.4 Commutative Algebras, Associative Algebras, and Division Algebras

The terminology introduced in chapter 5 to designate certain properties of hypercomplex systems carries over without changes to algebras. Thus an algebra is said to be *commutative* if for any two of its elements $\boldsymbol{a}$ and $\boldsymbol{b}$ we have

$$\boldsymbol{ab} = \boldsymbol{ba},$$

and *associative* if for any three of its elements $\boldsymbol{a}, \boldsymbol{b}, \boldsymbol{c}$ we have

$$(\boldsymbol{ab})\boldsymbol{c} = \boldsymbol{a}(\boldsymbol{bc}).$$

(It is of interest to note that early in the development of the theory of algebras the property of associativity seemed so natural that the term "algebra" meant "associative algebra".) Finally, an algebra $\mathcal{A}$ is called a *division algebra* if each of the equations

$$\boldsymbol{ax} = \boldsymbol{b} \tag{7.7}$$

and

$$\boldsymbol{ya} = \boldsymbol{b}, \tag{7.8}$$

where $\boldsymbol{a}$ and $\boldsymbol{b}$ are any elements of $\mathcal{A}$ and $\boldsymbol{a} \neq \mathbf{o}$, is uniquely solvable. In that case, the element $\boldsymbol{x}$ satisfying (7.7) is called the left quotient of $\boldsymbol{b}$ by $\boldsymbol{a}$, and the element $\boldsymbol{y}$ satisfying (7.8) is called the right quotient of $\boldsymbol{b}$ by $\boldsymbol{a}$.

It is easy to see that division algebras have the property that *if a product* $\boldsymbol{ab}$ *is zero, then either* $\boldsymbol{a}$ *or* $\boldsymbol{b}$ *is zero.* In fact, if $\boldsymbol{a} \neq \mathbf{o}$ then $\boldsymbol{b} = \mathbf{o}$, for $\mathbf{o}$ is a solution of $\boldsymbol{ax} = \mathbf{o}$ and such solutions are unique.

In chapter 9 we shall prove the converse proposition: *If* $\mathcal{A}$ *is an algebra such that* $\boldsymbol{ab} = \mathbf{o}$ *implies that either* $\boldsymbol{a}$ *or* $\boldsymbol{b}$ *is zero, then* $\mathcal{A}$ *is a division algebra.*

If $\boldsymbol{e}$ is an element of an algebra $\mathcal{A}$ such that

$$\boldsymbol{ae} = \boldsymbol{a} \text{ and } \boldsymbol{ea} = \boldsymbol{a}$$

for all $\boldsymbol{a} \in \mathcal{A}$, then $\boldsymbol{e}$ is called an identity of $\mathcal{A}$, and $\mathcal{A}$ is said to be *an algebra with an identity* element. As noted earlier, all hypercomplex systems are algebras with identities.

The simplest algebra with an identity is the one-dimensional algebra with the multiplication table

$$\boldsymbol{i}_1\boldsymbol{i}_1 = \boldsymbol{i}_1.$$

In this algebra the multiplication rule is

$$(a_1\boldsymbol{i}_1)(b_1\boldsymbol{i}_1) = a_1b_1\boldsymbol{i}_1.$$

Effectively, this rule coincides with the multiplication of real numbers. That is why we shall call this algebra the *algebra of real numbers.*

7.5 Examples

We consider examples of algebras that are not hypercomplex systems.
Example 7.1 *The n-dimensional null algebra.* The multiplication table of this algebra has a particularly simple form:

$$\boldsymbol{i}_\alpha \boldsymbol{i}_\beta = \mathbf{o}$$

for all α, β from 1 to n. It follows that the product of any two elements is zero.

Example 7.2 We consider the 2-dimensional algebra with the multiplication table

$$\begin{aligned} \boldsymbol{i}_1 \boldsymbol{i}_1 &= \boldsymbol{i}_1, \\ \boldsymbol{i}_1 \boldsymbol{i}_2 &= \boldsymbol{i}_2, \\ \boldsymbol{i}_2 \boldsymbol{i}_1 &= -\boldsymbol{i}_2, \\ \boldsymbol{i}_2 \boldsymbol{i}_2 &= \boldsymbol{i}_1. \end{aligned}$$

Here the multiplication rule is

$$(a_1 \boldsymbol{i}_1 + a_2 \boldsymbol{i}_2)(b_1 \boldsymbol{i}_1 + b_2 \boldsymbol{i}_2) = (a_1 b_1 + a_2 b_2)\boldsymbol{i}_1 + (a_1 b_2 - a_2 b_1)\boldsymbol{i}_2.$$

Notwithstanding the similarity between this multiplication and the multiplication of complex numbers, this algebra is different from the algebra of complex numbers. We leave it to the reader to show that this algebra has no identity element (and so cannot be a hypercomplex system). A more difficult exercise is proving the interesting fact that this algebra is a division algebra.

Example 7.3 *The algebra of 3-dimensional vectors with the cross product multiplication.* This algebra consists of elements of the form

$$b\boldsymbol{i} + c\boldsymbol{j} + d\boldsymbol{k},$$

multiplied in accordance with the table

$$\begin{gathered} \boldsymbol{i}^2 = \mathbf{o}, \quad \boldsymbol{j}^2 = \mathbf{o}, \quad \boldsymbol{k}^2 = \mathbf{o}, \\ \boldsymbol{ij} = \boldsymbol{k}, \quad \boldsymbol{ji} = -\boldsymbol{k}, \\ \boldsymbol{jk} = \boldsymbol{i}, \quad \boldsymbol{kj} = -\boldsymbol{i}, \\ \boldsymbol{ki} = \boldsymbol{j}, \quad \boldsymbol{ik} = -\boldsymbol{j}. \end{gathered}$$

Hence

$$\begin{aligned} &(b\boldsymbol{i} + c\boldsymbol{j} + d\boldsymbol{k})(b'\boldsymbol{i} + c'\boldsymbol{j} + d'\boldsymbol{k}) \\ &\quad = (cd' - dc')\boldsymbol{i} + (db' - bd')\boldsymbol{j} + (bc' - cb')\boldsymbol{k}. \end{aligned} \tag{7.9}$$

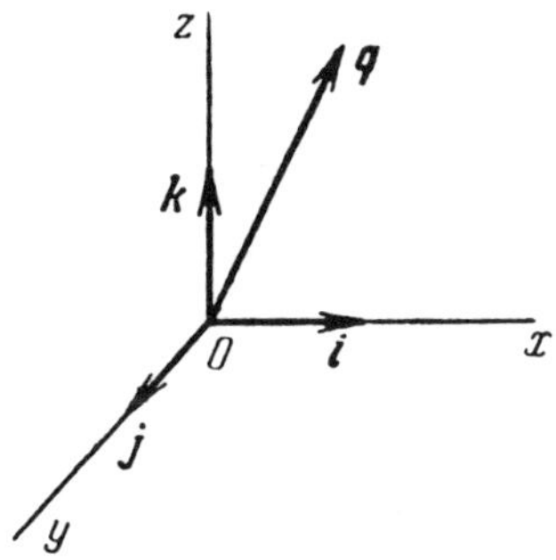

Figure 7.1.

▷ Consider a rectangular coordinate system in ordinary euclidean space. Let i, j, k be three unit vectors whose directions coincide with those of the coordinate axes (Figure 7.1). With the usual conventions for the sum of two vectors and the product of a number and a vector, the expression

$$q = b\boldsymbol{i} + c\boldsymbol{j} + d\boldsymbol{k}$$

represents a vector in space. The operation (7.9) is called the cross product (its geometric sense is discussed in chapter 4). It plays an important role in geometry and in physics. ◁

7.6 An Important Example: The Algebra of $n \times n$ Matrices

The dimension of this algebra is n^2. We could introduce n^2 imaginary units $\boldsymbol{i}_1, \boldsymbol{i}_2, \ldots, \boldsymbol{i}_{n^2}$, but it is more convenient to use another numbering scheme in which, instead of a number α taking on the values from 1 to n^2, we use the "number" α, β where α and β take on independently the values from 1 to n (so that the number of pairs α and β is n^2). Hence the notation $\boldsymbol{i}_{\alpha\beta}$ for an imaginary unit. The imaginary units can be ordered any way we wish. We choose the following ordering:

$$\boldsymbol{i}_{11}, \boldsymbol{i}_{12}, \ldots, \boldsymbol{i}_{1n} \vdots\ \boldsymbol{i}_{21}, \boldsymbol{i}_{22}, \ldots, \boldsymbol{i}_{2n} \vdots\ \boldsymbol{i}_{31}, \boldsymbol{i}_{32}, \ldots, \boldsymbol{i}_{3n} \vdots\ \ldots \vdots\ \boldsymbol{i}_{n1}, \boldsymbol{i}_{n2}, \ldots, \boldsymbol{i}_{nn}.$$

Thus the elements of our algebra are given by expressions of the form

$$\begin{aligned} A \;=\;& a_{11}\boldsymbol{i}_{11} + a_{12}\boldsymbol{i}_{12} + \ldots + a_{1n}\boldsymbol{i}_{1n} \\ +\;& a_{21}\boldsymbol{i}_{21} + a_{22}\boldsymbol{i}_{22} + \ldots + a_{2n}\boldsymbol{i}_{2n} \\ & \cdots\cdots\cdots\cdots\cdots\cdots\cdots\cdots \\ +\;& a_{n1}\boldsymbol{i}_{n1} + a_{n2}\boldsymbol{i}_{n2} + \ldots + a_{nn}\boldsymbol{i}_{nn}. \end{aligned} \tag{7.10}$$

This notation emphasizes the fact that each element A of our algebra is determined by a table

$$\begin{pmatrix} a_{11} & a_{12} & \ldots & a_{1n} \\ a_{21} & a_{22} & \ldots & a_{2n} \\ \ldots & \ldots & \ldots & \ldots \\ a_{n1} & a_{n2} & \ldots & a_{nn} \end{pmatrix}$$

with n^2 entries. Such tables are called *matrices*; more specifically, $n \times n$ *matrices*, or *square matrices of order n* (the order is the number of rows or columns of the square table). In what follows, we shall use the compact notation

$$A = \begin{pmatrix} a_{11} & a_{12} & \ldots & a_{1n} \\ a_{21} & a_{22} & \ldots & a_{2n} \\ \ldots & \ldots & \ldots & \ldots \\ a_{n1} & a_{n2} & \ldots & a_{nn} \end{pmatrix}$$

and suppose that *the elements of our algebra are matrices.*

Next we prescribe the multiplication table of the units $\boldsymbol{i}_{\alpha\beta}$ - the key to the "personality" of our matrix algebra. We put

$$\boldsymbol{i}_{\alpha 1}\boldsymbol{i}_{1\beta} = \boldsymbol{i}_{\alpha\beta},\; \boldsymbol{i}_{\alpha 2}\boldsymbol{i}_{2\beta} = \boldsymbol{i}_{\alpha\beta}, \ldots, \boldsymbol{i}_{\alpha n}\boldsymbol{i}_{n\beta} = \boldsymbol{i}_{\alpha\beta} \tag{7.10}$$

or, briefly,

$$\boldsymbol{i}_{\alpha\lambda}\boldsymbol{i}_{\lambda\beta} = \boldsymbol{i}_{\alpha\beta}.$$

Here α, β, λ range over the integers $1, \ldots, n$. By definition, all the remaining products of imaginary units are zero. The two "halves" of our definition can be combined in the single rule

$$\boldsymbol{i}_{\alpha\lambda}\boldsymbol{i}_{\mu\beta} = \delta_{\lambda\mu}\boldsymbol{i}_{\alpha\beta}, \tag{7.11}$$

where $\delta_{\lambda\mu}$ is defined to be 1 if $\lambda = \mu$ and 0 if $\lambda \neq \mu$.

We shall now try to determine the entries in the product of two elements A and B of our algebra, that is, to compute

$$\begin{pmatrix} & a_{11}\boldsymbol{i}_{11} + \ldots + a_{1n}\boldsymbol{i}_{1n} \\ + & a_{21}\boldsymbol{i}_{21} + \ldots + a_{2n}\boldsymbol{i}_{2n} \\ & \cdots\cdots\cdots\cdots\cdots \\ + & a_{n1}\boldsymbol{i}_{n1} + \ldots + a_{nn}\boldsymbol{i}_{nn} \end{pmatrix} \begin{pmatrix} & b_{11}\boldsymbol{i}_{11} + \ldots + b_{1n}\boldsymbol{i}_{1n} \\ + & b_{21}\boldsymbol{i}_{21} + \ldots + b_{2n}\boldsymbol{i}_{2n} \\ & \cdots\cdots\cdots\cdots\cdots \\ + & b_{n1}\boldsymbol{i}_{n1} + \ldots + b_{nn}\boldsymbol{i}_{nn} \end{pmatrix}.$$

The result is some element C:

$$C = \left(\begin{array}{rc} & c_{11}\boldsymbol{i}_{11} + c_{12}\boldsymbol{i}_{12} + \ldots + c_{1n}\boldsymbol{i}_{1n} \\ + & c_{21}\boldsymbol{i}_{21} + c_{22}\boldsymbol{i}_{22} + \ldots + c_{2n}\boldsymbol{i}_{2n} \\ & \ldots\ldots\ldots\ldots\ldots\ldots\ldots\ldots \\ + & c_{n1}\boldsymbol{i}_{n1} + c_{n2}\boldsymbol{i}_{n2} + \ldots + c_{nn}\boldsymbol{i}_{nn} \end{array} \right).$$

Consider a typical summand $c_{\alpha\beta}\boldsymbol{i}_{\alpha\beta}$ in C. The multiplication table (7.10) shows that this summand involves only the products

$$a_{\alpha 1}\boldsymbol{i}_{\alpha 1} \text{ by } b_{1\beta}\boldsymbol{i}_{1\beta},\ a_{\alpha 2}\boldsymbol{i}_{\alpha 2} \text{ by } b_{2\beta}\boldsymbol{i}_{2\beta}, \ldots, a_{\alpha n}\boldsymbol{i}_{\alpha n} \text{ by } b_{n\beta}\boldsymbol{i}_{n\beta},$$

and so is equal to

$$(a_{\alpha 1}b_{1\beta} + a_{\alpha 2}b_{2\beta} + \ldots + a_{\alpha n}b_{n\beta})\boldsymbol{i}_{\alpha\beta}.$$

In other words,

$$c_{\alpha\beta} = a_{\alpha 1}b_{1\beta} + a_{\alpha 2}b_{2\beta} + \ldots + a_{\alpha n}b_{n\beta}.$$

It is not difficult to remember this formula: *To obtain the element $c_{\alpha\beta}$ of the matrix C take the α-th row*

$$a_{\alpha 1} \quad a_{\alpha 2} \quad \cdots \quad a_{\alpha n}$$

of the matrix A and the β-th column

$$\begin{array}{c} b_{1\beta} \\ b_{2\beta} \\ \cdots \\ b_{n\beta} \end{array}$$

of the matrix B, form the product of each row element by the corresponding column element, and sum these products. The result is the element $c_{\alpha\beta}$ of the matrix C.

Given two matrices A and B, this rule enables us to compute a third matrix C which it is natural to call the product of the matrices A and B. In other words, to multiply two elements A and B of our algebra we must multiply their corresponding matrices A and B. For example, if

$$A = \begin{pmatrix} 1 & 2 \\ 3 & 4 \end{pmatrix} \quad \text{and} \quad B = \begin{pmatrix} 6 & -4 \\ -3 & 3 \end{pmatrix},$$

then

$$AB = \begin{pmatrix} 1\cdot 6 + 2\cdot(-3) & 1\cdot(-4) + 2\cdot 3 \\ 3\cdot 6 + 4\cdot(-3) & 3\cdot(-4) + 4\cdot 3 \end{pmatrix} = \begin{pmatrix} 0 & 2 \\ 6 & 0 \end{pmatrix}.$$

Matrix multiplication plays an extremely important role in mathematics. The modest scope of this book rules out a detailed study of this operation. We limit ourselves to proving that *matrix multiplication is associative*, or to put it differently, *the algebra of matrices is associative.*

For proof it suffices to show that the equality $(AB)C = A(BC)$ holds if for A, B, C we take arbitrary imaginary units of our matrix algebra (compare the similar argument used to prove the associativity of quaternion multiplication in chapter 3). In other words, we must prove that

$$(\boldsymbol{i}_{\alpha\lambda}\boldsymbol{i}_{\mu\beta})\boldsymbol{i}_{\nu\gamma} = \boldsymbol{i}_{\alpha\lambda}(\boldsymbol{i}_{\mu\beta}\boldsymbol{i}_{\nu\gamma}).$$

Now by (7.11), the left side is equal to $(\delta_{\lambda\mu}\boldsymbol{i}_{\alpha\beta})\boldsymbol{i}_{\nu\gamma}$, or, again by (7.11), to $\delta_{\lambda\mu}\delta_{\beta\nu}\boldsymbol{i}_{\alpha\gamma}$. Similarly, the right side is equal to $\boldsymbol{i}_{\alpha\lambda}(\delta_{\beta\nu}\boldsymbol{i}_{\mu\gamma})$, or to $\delta_{\beta\nu}\delta_{\lambda\mu}\boldsymbol{i}_{\alpha\gamma}$. Clearly, the two outcomes are equal.

7.7 Characterization of Multiplication in an Arbitrary Algebra

The material in this section is of an auxiliary nature and will be used only in chapters 16 and 18.

The following properties of the operation of multiplication are direct consequences of the definition of an algebra:

1. $(\boldsymbol{a} + \boldsymbol{b})\boldsymbol{c} = \boldsymbol{ac} + \boldsymbol{bc}, \quad \boldsymbol{a}(\boldsymbol{b} + \boldsymbol{c}) = \boldsymbol{ab} + \boldsymbol{ac},$

2. $k\boldsymbol{a} \cdot \boldsymbol{b} = k(\boldsymbol{ab}), \quad \boldsymbol{a} \cdot k\boldsymbol{b} = k(\boldsymbol{ab}).$

These properties *characterize* the multiplication operation in the sense clarified by the following proposition.

Consider the set $\mathcal{A}$ *of expressions of the form*

$$a_1\boldsymbol{i}_1 + a_2\boldsymbol{i}_2 + \ldots + a_n\boldsymbol{i}_n$$

with the operation (7.3) *of multiplication by a number, the operation* (7.4) *of addition, and a certain operation* $\boldsymbol{a} \circ \boldsymbol{b}$ *having the properties* 1 *and* 2 *above, that is*

$$\begin{aligned} (\boldsymbol{a} + \boldsymbol{b}) \circ \boldsymbol{c} = \boldsymbol{a} \circ \boldsymbol{c} + \boldsymbol{b} \circ \boldsymbol{c}, &\quad \boldsymbol{a} \circ (\boldsymbol{b} + \boldsymbol{c}) = \boldsymbol{a} \circ \boldsymbol{b} + \boldsymbol{a} \circ \boldsymbol{c}, \\ k\boldsymbol{a} \circ \boldsymbol{b} = k(\boldsymbol{a} \circ \boldsymbol{b}), &\quad \boldsymbol{a} \circ k\boldsymbol{b} = k(\boldsymbol{a} \circ \boldsymbol{b}). \end{aligned}$$

Then the set $\mathcal{A}$ *is an algebra whose multiplication operation is* $\boldsymbol{a} \circ \boldsymbol{b}$.

To prove this result we must check that $\boldsymbol{a} \circ \boldsymbol{b}$ is a multiplication in the sense of the definition of an algebra given in section 7.2.

Consider the expression $\boldsymbol{i}_\alpha \circ \boldsymbol{i}_\beta$. This is a certain element of $\mathcal{A}$, that is,

$$\boldsymbol{i}_\alpha \circ \boldsymbol{i}_\beta = p_{\alpha\beta,1}\boldsymbol{i}_1 + p_{\alpha\beta,2}\boldsymbol{i}_2 + \cdots + p_{\alpha\beta,n}\boldsymbol{i}_n. \tag{7.12}$$

Properties 1 and 2 imply that

$$\begin{aligned} \boldsymbol{a} \circ \boldsymbol{b} &= (a_1\boldsymbol{i}_1 + \ldots + a_n\boldsymbol{i}_n) \circ (b_1\boldsymbol{i}_1 + \ldots + b_n\boldsymbol{i}_n) \\ &\overset{!}{=} \sum_{\alpha,\beta} [(a_\alpha\boldsymbol{i}_\alpha) \circ (b_\beta\boldsymbol{i}_\beta)] \overset{!!}{=} \sum_{\alpha,\beta} a_\alpha b_\beta(\boldsymbol{i}_\alpha \circ \boldsymbol{i}_\beta) \end{aligned}$$

(here the equality marked with ! is justified by property 1 and the equality marked !! is justified by property 2). At this point, all we need do to compute $\boldsymbol{a} \circ \boldsymbol{b}$ is replace $\boldsymbol{i}_\alpha \circ \boldsymbol{i}_\beta$ by the corresponding element (7.12), multiply $a_\alpha b_\beta$ by this element, and carry out reductions. But this is just the procedure used in defining the multiplication of the elements in any algebra $\mathcal{A}$.

Part II

N-Dimensional Vectors

We turn once more to the definition of an n-dimensional algebra. Its most difficult component is, undoubtedly, the operation of multiplication. Is there anything significant left if this operation is suppressed? Well, what is left is a collection of elements that are uniquely represented in the form

$$a_1\boldsymbol{i}_1 + a_2\boldsymbol{i}_2 + \ldots + a_n\boldsymbol{i}_n,$$

with a natural rule of addition

$$\begin{aligned}(a_1\boldsymbol{i}_1 + a_2\boldsymbol{i}_2 + \ldots + a_n\boldsymbol{i}_n) + (b_1\boldsymbol{i}_1 + b_2\boldsymbol{i}_2 + \ldots + b_n\boldsymbol{i}_n)\\ = (a_1 + b_1)\boldsymbol{i}_1 + (a_2 + b_2)\boldsymbol{i}_2 + \ldots + (a_n + b_n)\boldsymbol{i}_n\end{aligned}$$

and an equally natural rule of multiplication of an element by a real number:

$$k(a_1\boldsymbol{i}_1 + a_2\boldsymbol{i}_2 + \ldots + a_n\boldsymbol{i}_n) = ka_1\boldsymbol{i}_1 + ka_2\boldsymbol{i}_2 + \ldots + ka_n\boldsymbol{i}_n.$$

While the material at our disposal does not look very promising, it turns out that it is possible to use it as the basis for a comprehensive theory. In fact, there is a whole branch of mathematics, called *linear algebra*, built around these two operations. Linear algebra has a rich content and is frequently utilized both is mathematics and in its many areas of application. The present chapter is an introduction to some aspects of linear algebra and will provide the basis for the study of the theory of algebras in part 3.

Chapter 8

The N-Dimensional Vector Space A_n

8.1 Basic Definitions

Definition 8.1 *By an n-*dimensional vector *we mean an object of the form*

$$a_1\boldsymbol{i}_1 + a_2\boldsymbol{i}_2 + \ldots + a_n\boldsymbol{i}_n, \tag{8.1}$$

where $a_1, a_2, \ldots, a_n$ are arbitrary real numbers and

$$\boldsymbol{i}_1, \boldsymbol{i}_2, \ldots, \boldsymbol{i}_n$$

are n different symbols to which we assign no special meaning.

We shall explain the reason for calling the expressions (8.1) vectors. If $n = 2$, then (8.1) reduces to

$$a_1\boldsymbol{i}_1 + a_2\boldsymbol{i}_2. \tag{8.2}$$

If we think of $\boldsymbol{i}_1$ and $\boldsymbol{i}_2$ as two fixed vectors in a plane, then the expression in (8.2) is again a vector in that plane[2] (Figure 8.1). Also, if the vectors $\boldsymbol{i}_1$ and $\boldsymbol{i}_2$ are not collinear, then every vector in the plane can be uniquely represented in the form (8.2).

Definition 8.2 *Two n-dimensional vectors*

$$a_1\boldsymbol{i}_1 + a_2\boldsymbol{i}_2 + \ldots + a_n\boldsymbol{i}_n$$

and

$$b_1\boldsymbol{i}_1 + b_2\boldsymbol{i}_2 + \ldots + b_n\boldsymbol{i}_n$$

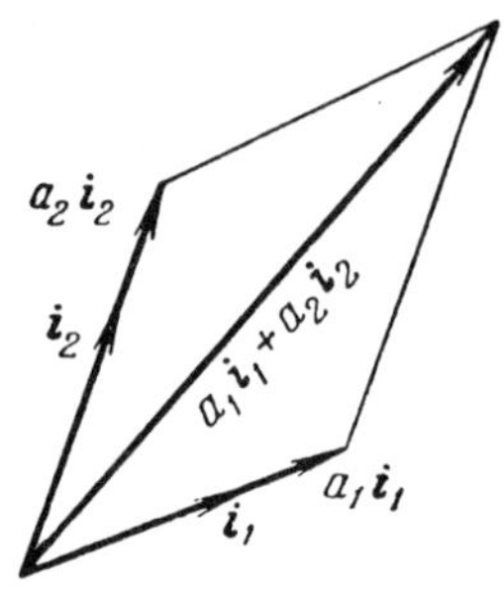

Figure 8.1.

are said to be equal *if and only if*

$$a_1 = b_1,\ a_2 = b_2, \ldots, a_n = b_n.$$

The reason for this definition was mentioned earlier: if $\boldsymbol{i}_1$ and $\boldsymbol{i}_2$ are two noncollinear "basis" vectors in the plane, then every vector in the plane has a unique representation (8.2).

Definition 8.3 *Two n-dimensional vectors are added according to the rule*

$$(a_1\boldsymbol{i}_1 + \ldots + a_n\boldsymbol{i}_n) + (b_1\boldsymbol{i}_1 + \ldots + b_n\boldsymbol{i}_n)$$
$$= (a_1 + b_1)\boldsymbol{i}_1 + \ldots + (a_n + b_n)\boldsymbol{i}_n,$$

and multiplication of an n-dimensional vector by a real number is determined by the rule

$$k(a_1\boldsymbol{i}_1 + a_2\boldsymbol{i}_2 + \ldots + a_n\boldsymbol{i}_n) = ka_1\boldsymbol{i}_1 + ka_2\boldsymbol{i}_2 + \ldots + ka_n\boldsymbol{i}_n.$$

This definition is also inspired by the corresponding definitions for geometric vectors.

We shall denote vectors briefly by boldface lowercase letters $\boldsymbol{a}$, $\boldsymbol{b}$, $\boldsymbol{c}$, and so on. Equality of vectors $\boldsymbol{a}$ and $\boldsymbol{b}$ will be denoted in the usual manner:

$$\boldsymbol{a} = \boldsymbol{b}.$$

It is easy to see that addition of vectors has the properties

$$\boldsymbol{a} + \boldsymbol{b} = \boldsymbol{b} + \boldsymbol{a} \qquad \text{(commutativity)},$$
$$(\boldsymbol{a} + \boldsymbol{b}) + \boldsymbol{c} = \boldsymbol{a} + (\boldsymbol{b} + \boldsymbol{c}) \qquad \text{(associativity)},$$

and multiplication of a vector by a scalar has the properties

$$k(l\boldsymbol{a}) = (kl)\boldsymbol{a},$$

$$(k+l)\boldsymbol{a} = k\boldsymbol{a} + l\boldsymbol{a}.$$

We shall call our system of vectors an *n-dimensional vector space* and denote it by A_n.

The vector

$$0\boldsymbol{i}_1 + 0\boldsymbol{i}_2 + \ldots + 0\boldsymbol{i}_n$$

is called the *zero vector* and is denoted by $\mathbf{o}$. It is clear that for any vector $\boldsymbol{a}$

$$\boldsymbol{a} + \mathbf{o} = \boldsymbol{a}$$

and

$$0\boldsymbol{a} = \mathbf{o}.$$

8.2 The Concept of Linear Dependence

When studying some problems we usually deal not with a single vector but with a whole system of n-dimensional vectors. Then we usually denote them by the same letter (say, $\boldsymbol{a}$) with different subscripts. Thus

$$\begin{aligned} \boldsymbol{a}_1 &= -5\boldsymbol{i}_1 + 3\boldsymbol{i}_2 + 5\boldsymbol{i}_3 + 3\boldsymbol{i}_4, \\ \boldsymbol{a}_2 &= -\boldsymbol{i}_1 + \boldsymbol{i}_2 + 4\boldsymbol{i}_3 + 3\boldsymbol{i}_4, \\ \boldsymbol{a}_3 &= \boldsymbol{i}_1 + 0\boldsymbol{i}_2 + 3\boldsymbol{i}_3 - 2\boldsymbol{i}_4, \end{aligned} \tag{8.3}$$

is an example of a system of three 4-dimensional vectors.

Let

$$\boldsymbol{a}_1, \boldsymbol{a}_2, \ldots, \boldsymbol{a}_m$$

be a system of n-dimensional vectors. We take arbitrary numbers

$$k_1, k_2, \ldots, k_m$$

and form the vector

$$\boldsymbol{a} = k_1\boldsymbol{a}_1 + k_2\boldsymbol{a}_2 + \ldots + k_m\boldsymbol{a}_m.$$

We say that the vector $\boldsymbol{a}$ is a *linear combination* of the vectors $\boldsymbol{a}_1, \boldsymbol{a}_2, \ldots, \boldsymbol{a}_m$ with coefficients $k_1, k_2, \ldots, k_m$.

Example 8.1 *Express the linear combination*

$$\boldsymbol{a}_1 - 3\boldsymbol{a}_2 + 2\boldsymbol{a}_3$$

of the vectors $\boldsymbol{a}_1, \boldsymbol{a}_2, \boldsymbol{a}_3$ *in* (8.3) *as a vector of the form* $k_1\boldsymbol{i}_1 + k_2\boldsymbol{i}_2 + k_3\boldsymbol{i}_3 + k_4\boldsymbol{i}_4$.

Solution 8.1 *Since*

$$\begin{array}{rcrcrcrcr} \boldsymbol{a}_1 & = & -5\boldsymbol{i}_1 & + & 3\boldsymbol{i}_2 & + & 5\boldsymbol{i}_3 & + & 3\boldsymbol{i}_4, \\ -3\boldsymbol{a}_2 & = & 3\boldsymbol{i}_1 & - & 3\boldsymbol{i}_2 & - & 12\boldsymbol{i}_3 & - & 9\boldsymbol{i}_4, \\ 2\boldsymbol{a}_3 & = & 2\boldsymbol{i}_1 & + & 0\boldsymbol{i}_2 & + & 6\boldsymbol{i}_3 & - & 4\boldsymbol{i}_4, \end{array}$$

it follows that

$$\boldsymbol{a}_1 - 3\boldsymbol{a}_2 + 2\boldsymbol{a}_3 = 0\boldsymbol{i}_1 + 0\boldsymbol{i}_2 - 1\boldsymbol{i}_3 - 10\boldsymbol{i}_4.$$

We introduce one more key definition.

Definition 8.4 *A system of vectors*

$$\boldsymbol{a}_1, \boldsymbol{a}_2, \ldots, \boldsymbol{a}_p \tag{8.4}$$

is said to be linearly dependent *if some linear combination of these vectors is equal to the zero vector,*

$$s_1\boldsymbol{a}_1 + s_2\boldsymbol{a}_2 + \ldots + s_p\boldsymbol{a}_p = \mathbf{o}, \tag{8.5}$$

and at least one of the coefficients $s_1, s_2, \ldots, s_p$ *is not zero. In the opposite case (that is, if no such linear combination exists) we say that the system* (8.4) *is* linearly independent.

A direct consequence of this definition is that a system consisting of a single vector is linearly dependent if the vector in question is the zero vector (indeed, if $s_1\boldsymbol{a}_1 = \mathbf{o}$ and $s_1 \neq 0$, then $\boldsymbol{a}_1 = \mathbf{o}$).

In case of a system of two vectors, linear dependence means that there are numbers s_1, s_2, not both zero, such that

$$s_1\boldsymbol{a}_1 + s_2\boldsymbol{a}_2 = \mathbf{o}.$$

Suppose that $s_1 \neq 0$. Then our equality implies that

$$\boldsymbol{a}_1 = k\boldsymbol{a}_2,$$

where $k = -s_2/s_1$. Two vectors so related are said to be *proportional.*

Now consider a system of p linearly dependent vectors and suppose, for definiteness, that s_1 in (8.5) is different from zero. Then

$$\boldsymbol{a}_1 = k_2\boldsymbol{a}_2 + k_3\boldsymbol{a}_3 + \ldots + k_p\boldsymbol{a}_p,$$

that is, the vector $\boldsymbol{a}_1$ is a linear combination of the other vectors in the system.

This argument shows that a linearly dependent system consists of one vector (and then it is the zero vector), or one of its vectors is a linear combination of the others.

8.3 Another Definition of Linear Dependence

We shall find it convenient to use another formulation of the concept of linear dependence: *the system* (8.4) *is linearly dependent if and only if* $\boldsymbol{a}_1 = \mathbf{o}$ *or one of its vectors is a linear combination of its predecessors.*

We shall show that our two definitions of linear dependence are equivalent.

If $\boldsymbol{a}_1 = \mathbf{o}$, then the system is linearly dependent for, in that case, the linear combination $1 \cdot \boldsymbol{a}_1 + 0\boldsymbol{a}_2 + \ldots + 0\boldsymbol{a}_p$ is equal to the zero vector and the first coefficient is different from zero. If one of the vectors is a linear combination of its predecessors,

$$\boldsymbol{a}_i = k_1\boldsymbol{a}_1 + \ldots + k_{i-1}\boldsymbol{a}_{i-1},$$

then the system is linearly dependent, for the linear combination $k_1\boldsymbol{a}_1 + \ldots + k_{i-1}\boldsymbol{a}_{i-1} - 1\boldsymbol{a}_i + 0\boldsymbol{a}_{i+1} + \ldots + 0\boldsymbol{a}_p$ is zero (that is, the zero vector) and the coefficient of $\boldsymbol{a}_i$ is not zero.

Conversely, suppose that our system is dependent, that is, (8.5) holds and at least one of the coefficients $s_1, s_2, \ldots, s_p$ is not zero. Consider the last of the nonzero coefficients. If that is s_1, then $s_1\boldsymbol{a}_1 = \mathbf{o}$. But then $\boldsymbol{a}_1 = \mathbf{o}$. If that is s_i with $i > 1$ then, by adding the vector $-s_i\boldsymbol{a}_i$ to our equality and multiplying both sides by $-1/s_i$, we obtain an equality of the form

$$\boldsymbol{a}_i = k_1\boldsymbol{a}_1 + \ldots + k_{i-1}\boldsymbol{a}_{i-1},$$

that is, we shall have expressed the vector $\boldsymbol{a}_i$ as a linear combination of its predecessors.

8.4 The Initial Basis

It is natural to denote the vector

$$1\boldsymbol{i}_1 + 0\boldsymbol{i}_2 + \ldots + 0\boldsymbol{i}_n$$

as $\boldsymbol{i}_1$. This means that the symbol $\boldsymbol{i}_1$, originally devoid of any particular meaning, has now been identified with one of the vectors. Similarly, we identify

$$0\boldsymbol{i}_1 + 1\boldsymbol{i}_2 + \ldots + 0\boldsymbol{i}_n$$

with $\boldsymbol{i}_2$, and so on.

The vectors

$$\boldsymbol{i}_1, \boldsymbol{i}_2, \ldots, \boldsymbol{i}_n$$

just defined have the property that every vector in A_n can be expressed as a linear combination of them. In fact, consider any vector $\boldsymbol{a} \in \mathrm{A}_n$. This vector is some formal sum

$$a_1\boldsymbol{i}_1 + a_2\boldsymbol{i}_2 + \ldots + a_n\boldsymbol{i}_n. \tag{8.6}$$

In view of definitions 8.2 and 8.3 we can write

$$\begin{aligned} a_1\boldsymbol{i}_1 + a_2\boldsymbol{i}_2 + \ldots + a_n\boldsymbol{i}_n &= a_1(1\boldsymbol{i}_1 + 0\boldsymbol{i}_2 + \ldots + 0\boldsymbol{i}_n) \\ &+ a_2(0\boldsymbol{i}_1 + 1\boldsymbol{i}_2 + \ldots 0\boldsymbol{i}_n) + \ldots + a_n(0\boldsymbol{i}_1 + 0\boldsymbol{i}_2 + \ldots + 1\boldsymbol{i}_n), \end{aligned}$$

which means that the vector $\boldsymbol{a}$ is a linear combination of the vectors $\boldsymbol{i}_1, \boldsymbol{i}_2, \ldots, \boldsymbol{i}_n$ with the coefficients $a_1, a_2, \ldots, a_n$. This means that we may now regard the formal sum (8.6) as a genuine linear combination of vectors.

The vectors $\boldsymbol{i}_1, \boldsymbol{i}_2, \ldots, \boldsymbol{i}_n$ form a so-called *basis* of the space A_n. In the next section we shall give a precise definition of this term. By way of an anticipatory remark we wish to note that there are infinitely many bases, and that the basis $\boldsymbol{i}_1, \boldsymbol{i}_2, \ldots, \boldsymbol{i}_n$ is in no way distinguished.

Chapter 9

A Basis of The Space A_n

9.1 Definition of a Basis

A finite system of vectors

$$\boldsymbol{a}_1, \boldsymbol{a}_2, \ldots, \boldsymbol{a}_p \tag{9.1}$$

in called a *basis of the space* A_n if it has the following two properties:

1. every vector $\boldsymbol{a} \in \mathrm{A}_n$ is a linear combination of these vectors,

$$\boldsymbol{a} = k_1\boldsymbol{a}_1 + k_2\boldsymbol{a}_2 + \ldots + k_p\boldsymbol{a}_p; \tag{9.2}$$

2. the representation (9.2) of $\boldsymbol{a}$ is unique, that is, given a representation

$$\boldsymbol{a} = l_1\boldsymbol{a}_1 + l_2\boldsymbol{a}_2 + \ldots + l_p\boldsymbol{a}_p,$$

   we can conclude that

$$k_1 = l_1, \quad k_2 = l_2, \quad \ldots, \quad k_p = l_p.$$

We shall prove that *the vectors of a basis are linearly independent.*

Suppose that the vectors (9.1) are linearly dependent. By definition, there is a linear combination of these vectors that is equal to zero,

$$s_1\boldsymbol{a}_1 + s_2\boldsymbol{a}_2 + \ldots + s_p\boldsymbol{a}_p = \boldsymbol{o}, \tag{9.3}$$

and not all of the coefficients $s_1, s_2, \ldots, s_p$ are zero. By adding the equalities (9.2) and (9.3) we obtain

$$\boldsymbol{a} = (k_1 + s_1)\boldsymbol{a}_1 + (k_2 + s_2)\boldsymbol{a}_2 + \ldots + (k_p + s_p)\boldsymbol{a}_p,$$

that is, a *different* representation of $\boldsymbol{a}$ as a linear combination of the vectors (9.1). But this contradicts the definition of a basis.

9.2 Obtaining Other Bases

One example of a basis is the system

$$\boldsymbol{i}_1, \boldsymbol{i}_2, \ldots, \boldsymbol{i}_n$$

of initial basis vectors. It is obvious that these vectors satisfy the conditions 1 and 2 in section 9.1. This is far from being the only basis of the space A_n, however. What follows are two ways of obtaining bases from a given basis.

1. Multiply any one of the basis vectors by a nonzero number.

   For example, by multiplying the first of the vectors in (9.1) by some nonzero number k we obtain the system of vectors

   $$k\boldsymbol{a}_1, \boldsymbol{a}_2, \ldots, \boldsymbol{a}_p, \tag{9.1'}$$

   which is obviously a basis.

2. Replace one of the basis vectors by the sum of that vector and another of the basis vectors.

   For example, by replacing the vector $\boldsymbol{a}_1$ by the sum $\boldsymbol{a}_1 + \boldsymbol{a}_2$ we obtain the new system

   $$\boldsymbol{a}_1 + \boldsymbol{a}_2, \quad \boldsymbol{a}_2, \quad \ldots, \quad \boldsymbol{a}_p, \tag{9.1''}$$

   which is again a basis. In fact , let $\boldsymbol{a}$ be some vector. Then, for some real numbers $k_1, \ldots, k_p$, equality (9.2) holds. But then

   $$\boldsymbol{a} = k_1(\boldsymbol{a}_1 + \boldsymbol{a}_2) + (k_2 - k_1)\boldsymbol{a}_2 + k_3\boldsymbol{a}_3 + \ldots + k_p\boldsymbol{a}_p,$$

   that is, $\boldsymbol{a}$ is a linear combination of the vectors (9.1″). Also, the uniqueness of the representation of $\boldsymbol{a}$ in terms of the basis vectors (9.1) readily implies the uniqueness of its representation in terms of the vectors (9.1″). Hence (9.1″) is also a basis of the space A_n.

▷ It is natural to ask how one finds all bases of the space A_n. What we have in mind is a procedure for obtaining from any basis all the others. In a sense, the following proposition answers this question. In this proposition, the term "elementary transformations" refers to the above two ways of obtaining bases from a given basis.

It is possible to go from any basis to any other basis by means of a finite number of elementary transformations.

In particular, it is possible to obtain all bases of the space A_n by applying (arbitrary numbers of) all possible elementary transformations to the basis of the initial basis vectors.

The proof of this proposition is not difficult but we shall not give it here. Actually, it is not so much this proposition that interests us here but rather its consequence which asserts that:

Every basis of the space A_n *consists of* n *vectors.*

Given the above proposition, the proof of its consequence just formulated is obvious, for an elementary transformation preserves the number of vectors in any system of vectors.

Below we give an independent proof of the last proposition. ◁

9.3 The Number of Basis Vectors

We shall now prove the following theorem.

Theorem 9.1 *Every basis of the space* A_n *consists of* n *vectors.*

Since the initial basis $\boldsymbol{i}_1, \boldsymbol{i}_2, \ldots, \boldsymbol{i}_n$ consists of n vectors, all we need show is that *any two bases contain equal numbers of vectors.*

Before embarking on the proof we make the following observation.

Suppose that the system of vectors

$$\boldsymbol{a}_1, \boldsymbol{a}_2, \ldots, \boldsymbol{a}_p$$

is *complete*, by which we mean that any vector $\boldsymbol{a}$ can be written as a linear combination of these vectors. Take any nonzero vector $\boldsymbol{b}$ and adjoin it to our complete system as a new first vector. The new system of vectors

$$\boldsymbol{b}, \boldsymbol{a}_1, \boldsymbol{a}_2, \ldots, \boldsymbol{a}_p$$

is linearly dependent (for the completeness of the initial system implies an equality of the form $\boldsymbol{b} - k_1\boldsymbol{a}_1 - k_2\boldsymbol{a}_2 - \ldots - k_p\boldsymbol{a}_p = \mathbf{o}$). According to section 3 of chapter 8, the new system contains a vector $\boldsymbol{a}_i$ that can be written as a linear combination of its predecessors. We claim that if we eliminate the "superfluous" vector $\boldsymbol{a}_i$, then *the resulting system is again complete.*

This is almost obvious. Indeed, any vector $\boldsymbol{a}$ can be written as a linear combination of the vectors $\boldsymbol{a}_1, \boldsymbol{a}_2, \ldots, \boldsymbol{a}_p$. If we replace in this expression the "superfluous" vector $\boldsymbol{a}_i$ by its representation in terms of a linear combination of its predecessors $\boldsymbol{b}, \boldsymbol{a}_1, \boldsymbol{a}_2, \ldots, \boldsymbol{a}_{i-1}$, then we shall have expressed the vector $\boldsymbol{a}$ as a linear combination of the p vectors $\boldsymbol{b}, \boldsymbol{a}_1, \boldsymbol{a}_2, \ldots, \boldsymbol{a}_{i-1}, \boldsymbol{a}_{i+1}, \ldots, \boldsymbol{a}_p$.

Now the proof of theorem 9.1 is quite short. Thus, let

$$\boldsymbol{a}_1, \boldsymbol{a}_2., \ldots, \boldsymbol{a}_p \tag{9.4}$$

and

$$\boldsymbol{b}_1, \boldsymbol{b}_2., \ldots, \boldsymbol{b}_q \tag{9.5}$$

be two bases. We wish to show that $p = q$.

Suppose that $p \neq q$ and assume, for definiteness, that $p < q$. Adjoin the vector $\boldsymbol{b}_1$ as a first vector to the system (9.4) and eliminate from it the "superfluous" vector. By what was proved above, the new system

$$\boldsymbol{b}_1, \underbrace{\ldots\ldots}_{p-1},$$

call it (9.4′), is complete.

Adjoin the vector $\boldsymbol{b}_2$ as a first vector to the system (9.4′) and eliminate from the resulting system the "superfluous" vector. This yields a new complete system

$$\boldsymbol{b}_2, \boldsymbol{b}_1, \underbrace{\ldots\ldots}_{p-2},$$

and so on.

Note that none of the adjoined vectors $\boldsymbol{b}_1, \boldsymbol{b}_2, \ldots$ can be a "superfluous" vector. This is so because these vectors belong to a basis (namely, (9.5)). This means that at each step of our process we eliminate one of the vectors $\boldsymbol{a}_1, \boldsymbol{a}_2, \ldots, \boldsymbol{a}_p$.

After p steps we will have eliminated all the vectors $\boldsymbol{a}_1, \boldsymbol{a}_2, \ldots, \boldsymbol{a}_p$ and arrived at the presumably complete system

$$\boldsymbol{b}_p, \boldsymbol{b}_{p-1}, \ldots, \boldsymbol{b}_2, \boldsymbol{b}_1.$$

But this is impossible, for the vector $\boldsymbol{b}_{p+1}$, say, cannot be written as a linear combination of these vectors. This contradiction proves our theorem.

9.4 The Number of Vectors in a Linearly Independent System

We proved above that the vectors of a basis are linearly independent. It follows trivially that the space A_n contains linearly independent systems of n vectors. It is natural to ask whether there are in A_n systems of *more* than n linearly independent vectors. It turns out that this is not the case. In fact, we have

Theorem 9.2 *If the system*

$$\boldsymbol{a}_1, \boldsymbol{a}_2, \ldots, \boldsymbol{a}_p$$

of vectors in A_n *is linearly independent, then* $p \leq n$. *If* $p = n$, *then the given system is a basis of the space* A_n.

Proof. Denote the given system for brevity by S. We shall construct a basis of the space A_n by adjoining to the system S vectors from an arbitrary basis

$$e_1, e_2, \ldots, e_n.$$

Consider the vector e_1. If it can be written as a linear combination of the vectors in S, then we ignore it. If not, then we adjoin it to S (as a last vector). In either case, we call the new system S' (we have $S' = S$ or $S' = S \cup e_1$).

Now we consider e_2. If it can be written as a linear combination of the vectors in S', then we ignore it. Otherwise we adjoin it to S'. In either case we call the new system S'' (we have $S'' = S'$ or $S'' = S' \cup e_2$). After n such steps we end up with a system $S^{(n)}$. This system has the following properties:

1. Every vector $a \in A_n$ is a linear combination of vectors in $S^{(n)}$. This is so because the vector a is a linear combination of the vectors $e_1, e_2, \ldots, e_n$ which, in turn, are linear combinations of vectors in $S^{(n)}$ (the latter follows from the way we constructed $S^{(n)}$).

2. None of the vectors in $S^{(n)}$ can be written as a linear combination of its predecessors (this again follows from the manner of construction of $S^{(n)}$). This means that the system $S^{(n)}$ is linearly independent.

3. The representation of any vector a as a linear combination of the vectors in $S^{(n)}$ is unique. Otherwise, the difference of two representations of a would yield a relation of linear dependence connecting vectors in $S^{(n)}$.

Properties 1 and 3 show that the system $S^{(n)}$ is a basis of the space A_n. By theorem 9.1, the number of vectors in $S^{(n)}$ is n. Since $S^{(n)}$ contains the vectors $a_1, a_2, \ldots, a_p$, it follows that $p \leq n$. If $p = n$, then our initial system is a basis. This completes the proof.

9.5 A Consequence of Theorem 9.2 Pertaining to Algebras

In chapter 7 we promised to prove that in an algebra $\mathcal{A}$ has the property that $ab = o$ implies $a = o$ or $b = o$, then $\mathcal{A}$ is a division algebra. We couldn't prove this proposition then but we prove it now.

Suppose that we wish to solve the equation

$$ax = b, \tag{9.6}$$

where $\boldsymbol{a} \neq \mathbf{o}$. We choose a basis

$$\boldsymbol{e}_1, \boldsymbol{e}_2, \ldots, \boldsymbol{e}_n$$

in the vector space $\mathcal{A}$. Upon multiplication of the basis vectors on the left by $\boldsymbol{a}$ we obtain the system of n vectors

$$\boldsymbol{a}\boldsymbol{e}_1, \boldsymbol{a}\boldsymbol{e}_2, \ldots, \boldsymbol{a}\boldsymbol{e}_n. \tag{9.7}$$

We prove that these vectors again form a basis.

By Theorem 9.2, it suffices to show that the vectors (9.7) are linearly independent. Suppose that this is false. Then there exist numbers $k_1, k_2, \ldots, k_n$ not all zero such that

$$k_1\boldsymbol{a}\boldsymbol{e}_1 + k_2\boldsymbol{a}\boldsymbol{e}_2 + \ldots + k_n\boldsymbol{a}\boldsymbol{e}_n = \mathbf{o}.$$

Hence

$$\boldsymbol{a}(k_1\boldsymbol{e}_1 + k_2\boldsymbol{e}_2 + \ldots + k_n\boldsymbol{e}_n) = \mathbf{o}.$$

By assumption, one of the factors in the last equation must be $\mathbf{o}$. Since $\boldsymbol{a} \neq \mathbf{o}$,it follows that

$$k_1\boldsymbol{e}_1 + k_2\boldsymbol{e}_2 + \ldots + k_n\boldsymbol{e}_n = \mathbf{o}.$$

This contradicts the linear independence of the vectors $\boldsymbol{e}_1, \boldsymbol{e}_2, \ldots, \boldsymbol{e}_n$.

Thus the vectors (9.7) form a basis. Write $\boldsymbol{b}$ as a linear combination of the vectors in (9.7),

$$\boldsymbol{b} = s_1\boldsymbol{a}\boldsymbol{e}_1 + s_2\boldsymbol{a}\boldsymbol{e}_2 + \ldots + s_n\boldsymbol{a}\boldsymbol{e}_n.$$

If we rewrite this as

$$\boldsymbol{b} = \boldsymbol{a}(s_1\boldsymbol{e}_1 + s_2\boldsymbol{e}_2 + \ldots + s_n\boldsymbol{e}_n),$$

then we see that the element

$$\boldsymbol{x} = s_1\boldsymbol{e}_1 + s_2\boldsymbol{e}_2 + \ldots + s_n\boldsymbol{e}_n$$

is a solution of (9.6). This solution is unique. In fact, if $\boldsymbol{x}'$ were another solution, then by forming the difference of the equations $\boldsymbol{a}\boldsymbol{x} = \boldsymbol{b}$ and $\boldsymbol{a}\boldsymbol{x}' = \boldsymbol{b}$ we would obtain

$$\boldsymbol{a}(\boldsymbol{x} - \boldsymbol{x}') = \mathbf{o},$$

which implies that $\boldsymbol{x} - \boldsymbol{x}' = \mathbf{o}$, that is, $\boldsymbol{x} = \boldsymbol{x}'$.

A similar argument proves the existence and uniqueness of the solution of the equation

$$\boldsymbol{x}\boldsymbol{a} = \boldsymbol{b}.$$

9.6 Coordinates of a Vector Relative to a Basis

The last question we want to touch on in this section is that of the coordinates of a vector relative to a given basis of the space A_n.

Thus, let

$$\boldsymbol{a}_1, \boldsymbol{a}_2, \ldots, \boldsymbol{a}_n$$

be a basis of the space A_n, and let

$$\boldsymbol{p} = k_1\boldsymbol{a}_1 + k_2\boldsymbol{a}_2 + \ldots + k_n\boldsymbol{a}_n$$

be the representation of a vector $\boldsymbol{p} \in A_n$ relative to this basis. The numbers $k_1, k_2, \ldots, k_n$ are called the *coordinates of the vector* $\boldsymbol{p}$ *relative to the given basis.*

The equality

$$(k_1\boldsymbol{a}_1 + \ldots + k_n\boldsymbol{a}_n) + (l_1\boldsymbol{a}_1 \ldots + l_n\boldsymbol{a}_n)$$
$$= (k_1 + l_1)\boldsymbol{a}_1 + \ldots + (k_n + l_n)\boldsymbol{a}_n,$$

implied by the properties of vector addition and multiplication of a vector by a number, shows that when vectors are added, then their corresponding coordinates are added. Similarly, the equality

$$k(k_1\boldsymbol{a}_1 + k_2\boldsymbol{a}_2 + \ldots + k_n\boldsymbol{a}_n) = kk_1\boldsymbol{a}_1 + kk_2\boldsymbol{a}_2 + \ldots + kk_n\boldsymbol{a}_n$$

shows that when a vector is multiplied by a number, then its coordinates are multiplied by that number. In other words, the rules of addition of vectors and multiplication of a vector by a number are the same for the initial basis $\boldsymbol{i}_1, \boldsymbol{i}_2, \ldots, \boldsymbol{i}_n$ as well as for any other basis $\boldsymbol{a}_1, \boldsymbol{a}_2, \ldots, \boldsymbol{a}_n$.

Chapter 10

Subspaces

There are certain subsets of the space A_n whose properties justify our regarding them as independent spaces $A_p, p \leq n$. We call such subsets *subspaces* of A_n.

10.1 Definition of a Subspace

Let P be a nonempty set of vectors in A_n. We shall call it a *subspace* of the space A_n if

1. $\boldsymbol{a} \in \mathrm{P}$ and $\boldsymbol{b} \in \mathrm{P}$ imply that $\boldsymbol{a} + \boldsymbol{b} \in \mathrm{P}$;

2. $\boldsymbol{a} \in \mathrm{P}$ implies that $k\boldsymbol{a} \in \mathrm{P}$ for any real number k.

In other words, a subspace is a set of vectors containing all linear combinations $k\boldsymbol{a} + l\boldsymbol{b} + s\boldsymbol{c} + \ldots$ of vectors $\boldsymbol{a}, \boldsymbol{b}, \boldsymbol{c}, \ldots$ in it.

Trivial examples of subspaces are the so-called *null* space, consisting of the zero vector, and the space A_n. But there are many other subspaces. In the next section we shall explain the structure of any subspace of A_n.

Suppose that P is a subspace that is not the null space. Let $\boldsymbol{a}_1 \in \mathrm{P}, \boldsymbol{a}_1 \neq \mathbf{o}$. If all the vectors of P are multiples of $\boldsymbol{a}_1$, then we are finished. Otherwise we adjoin to $\boldsymbol{a}_1$ a vector $\boldsymbol{a}_2$ in P that is not a multiple of $\boldsymbol{a}_1$. If all the vectors in P are linear combinations of $\boldsymbol{a}_1$ and $\boldsymbol{a}_2$, then we are finished. Otherwise we adjoin to $\boldsymbol{a}_1, \boldsymbol{a}_2$ a vector $\boldsymbol{a}_3$ in P that is not a linear combination of $\boldsymbol{a}_1$ and $\boldsymbol{a}_2$, and so on. This process yields a system of vectors none of which is a linear combination of its predecessors. This means that at each step we are dealing with a linearly independent system of vectors. By theorem 9.2 of the preceding chapter this process must end after at most n steps. In other words, we obtain a system of linearly independent vectors

$$\boldsymbol{a}_1, \boldsymbol{a}_2, \ldots, \boldsymbol{a}_p \ (p \leq n), \tag{10.1}$$

in P such that

1. every vector in P is a linear combination of these vectors, and
2. this linear combination is unique. (In fact, if we had two such linear combinations for a vector $\boldsymbol{a} \in P$, then their difference $s_1\boldsymbol{a}_1 + s_2\boldsymbol{a}_2 + \ldots + s_p\boldsymbol{a}_p$ would be equal to $\mathbf{o}$ and at least one of the coefficients $s_1, s_2, \ldots, s_p$ would be different from zero. But this would contradict the linear independence of the vectors (10.1).)

We see that the subspace P consists of all vectors of the form

$$k_1\boldsymbol{a}_1 + k_2\boldsymbol{a}_2 + \ldots + k_p\boldsymbol{a}_p$$

and that there is just one such representation for every vector $\boldsymbol{a} \in \mathbf{P}$. This allows us to regard P as a p-dimensional vector space A_p with initial basis $\boldsymbol{a}_1, \boldsymbol{a}_2, \ldots, \boldsymbol{a}_p$. Of course, the theorems proved earlier for vector spaces hold in this space. In particular, each of its bases consists of p vectors. The number p is called the ***dimension of the subspace*** **P**. We saw that p cannot exceed n. If $p = n$, then the system (10.1) is a basis of A_n (see once more theorem 9.2 of the preceding chapter), so that P coincides with A_n.

We wish to emphasize the following immediate consequence of the above: *every subspace* **P** *coincides with the totality of linear combinations of certain* p *vectors* $\boldsymbol{a}_1, \boldsymbol{a}_2, \ldots, \boldsymbol{a}_p$.

10.2 Examples

We illustrate the concept of a subspace in the case of the 3-dimensional space A_3.

Let P be a nonnull subspace of A_3. A basis of P contains at most three vectors. It follows that a basis of P has one of the following three forms:

$$\boldsymbol{a}_1; \quad \boldsymbol{a}_1, \boldsymbol{a}_2; \quad \boldsymbol{a}_1, \boldsymbol{a}_2, \boldsymbol{a}_3.$$

In the first case, P consists of all multiples $k\boldsymbol{a}_1$ of $\boldsymbol{a}_1$ (Figure 10.1). In the second case P is the set of all vectors of the form $k_1\boldsymbol{a}_1 + k_2\boldsymbol{a}_2$, and so consists of all vectors coplanar with $\boldsymbol{a}_1$ and $\boldsymbol{a}_2$ (Figure 10.2). In the third case, P consists of all vectors of the form $k_1\boldsymbol{a}_1 + k_2\boldsymbol{a}_2 + k_3\boldsymbol{a}_3$, that is, of all vectors in A_3 (Figure 10.3).

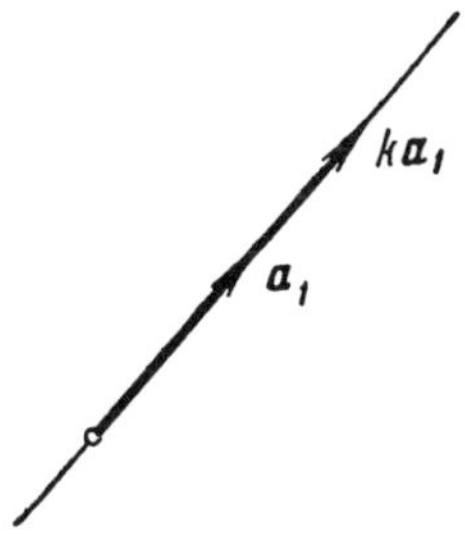

Figure 10.1.

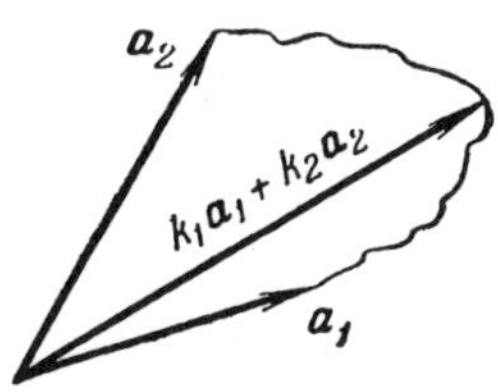

Figure 10.2.

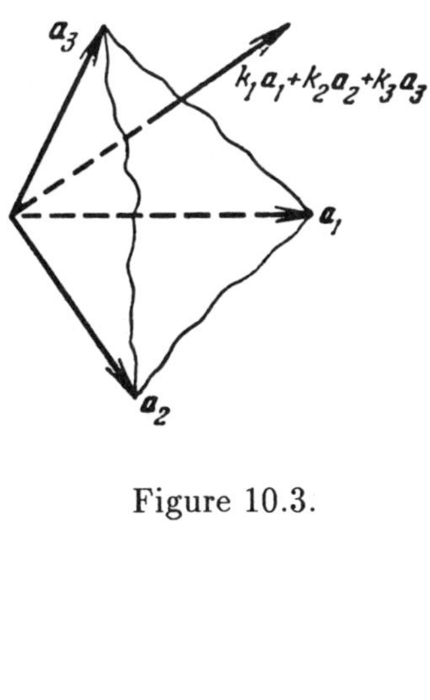

Figure 10.3.

Chapter 11

Lemma on Homogeneous Systems of Equations

This chapter is of an auxiliary nature. In it we consider a subject seemingly unrelated to vectors, namely, the subject of *systems of linear equations.* More accurately, we shall prove a lemma pertaining to systems of *homogeneous* linear equations. This lemma will help us establish important results to be considered in the sequel.

A linear equation is called homogeneous if its free term is zero. In other words, a homogeneous linear equation in n unknowns $x_1, x_2, \ldots, x_n$ has the form

$$a_1x_1 + a_2x_2 + \ldots + a_nx_n = 0.$$

A system consisting of homogeneous linear equations is itself called homogeneous. A homogeneous system of m equations in n unknowns has the form

$$\begin{array}{l} a_1x_1 + a_2x_2 + \ldots + a_nx_n = 0 - \text{ 1st equation,} \\ b_1x_1 + b_2x_2 + \ldots + b_nx_n = 0 - \text{ 2nd equation,} \\ \ldots\ldots\ldots\ldots\ldots\ldots\ldots\ldots\ldots\ldots\ldots\ldots \\ d_1x_1 + d_2x_2 + \ldots + d_nx_n = 0 - \text{ mth equation.} \end{array} \tag{11.1}$$

An obvious solution of a homogeneous system is the so-called *null* solution.

$$x_1 = 0, \quad x_2 = 0, \quad \ldots, \quad x_n = 0.$$

Frequently it is important to know if a homogeneous system has nonnull solutions. A partial answer to this question is provided by the following lemma.

Lemma 11.1 *A homogeneous system in which the number of equations is smaller than the number of unknowns always has a nonnull solution.*

The proof is by induction on the number m of equations in the system (11.1).

If $m = 1$, then we have a single equation in more than one unknown. It is clear that such an equation has a nonnull solution.

We assume that our lemma holds for systems with $m - 1$ equations and prove it for systems of m equations.

If the coefficients $a_1, b_1, \ldots, d_1$ of x_1 are all zero, then our system has nonnull solutions. An example of such a solution is

$$x_1 = 1, \quad x_2 = 0, \quad \ldots, \quad x_n = 0.$$

Now suppose that a_1 is not zero. (Note that this is a harmless assumption that may require, at most, the rearranging of the equations of our system.)

We transform our system as follows. We multiply the first equation by $-b_1/a_1$ and add it to the second equation. The result is a new system, equivalent to the first, whose second equation has the form

$$b'_2 x_2 + \ldots + b'_n x_n = 0$$

(the coefficient of x_1 is 0). By adding appropriate multiples of the first equation to the remaining equations of the system, if any, we end up with a homogeneous system of the form

$$a_1 x_1 + a_2 x_2 + \ldots + a_n x_n = 0,$$

$$\boxed{\begin{array}{l} b'_2 x_2 + \ldots + b'_n x_n = 0, \\ \ldots\ldots\ldots\ldots\ldots\ldots\ldots \\ d'_2 x_2 + \ldots + d'_n x_n = 0, \end{array}} \tag{11.1'}$$

that is equivalent to the starting system. The *boxed* part of the system (11.1′) is a homogeneous system of $m-1$ equations in $n-1$ unknowns. Since $m < n$,

$$m - 1 < n - 1,$$

so that the boxed system has fewer equations than unknowns. Also, the boxed system has $m-1$ equations. But in view of the induction assumption, the boxed system has a nonnull solution

$$x_2 = \alpha_2, \quad x_3 = \alpha_3, \quad \ldots, \quad x_n = \alpha_n.$$

By adjoining to it the value of x_1 obtained from the first equation of the system (11.1′), namely,

$$x_1 = \frac{1}{a_1}(a_2\alpha_2 + \ldots + a_n\alpha_n),$$

we obtain a nonnull solution of the system (11.1′), and thus of original system (11.1). This proves our lemma.

Chapter 12

Scalar Products

All the concepts studied thus far in this part are based on two operations on vectors: addition of vectors and multiplication of vector by a number. If we consider geometric vectors (that is, directed segments in the plane or in space), then there are many concepts, such as length of a vector, perpendicular vectors, and so on, for which we have so far not provided sensible analogues in A_n. This we do next.

12.1 The Scalar Product of Geometric Vectors

Let $\boldsymbol{x}$ and $\boldsymbol{y}$ be two vectors in the plane emanating from the origin O. Let the coordinates of $\boldsymbol{x}$ and $\boldsymbol{y}$ be, respectively, x_1, x_2 and y_1, y_2. Then

$$\begin{aligned} \boldsymbol{x} &= x_1\boldsymbol{i}_1 + x_2\boldsymbol{i}_2, \\ \boldsymbol{y} &= y_1\boldsymbol{i}_1 + y_2\boldsymbol{i}_2, \end{aligned}$$

where $\boldsymbol{i}_1, \boldsymbol{i}_2$ are unit vectors whose orientations are those of the coordinate axes (Figure 12.1).

Let X and Y be the endpoints of our vectors. Then the coordinates of X are x_1, x_2 and the coordinates of Y are y_1, y_2.

The formula for the distance between two points yields the relations

$$\begin{aligned} XY^2 &= (y_1 - x_1)^2 + (y_2 - x_2)^2, \\ OX^2 &= x_1^2 + x_2^2, \\ OY^2 &= y_1^2 + y_2^2, \end{aligned}$$

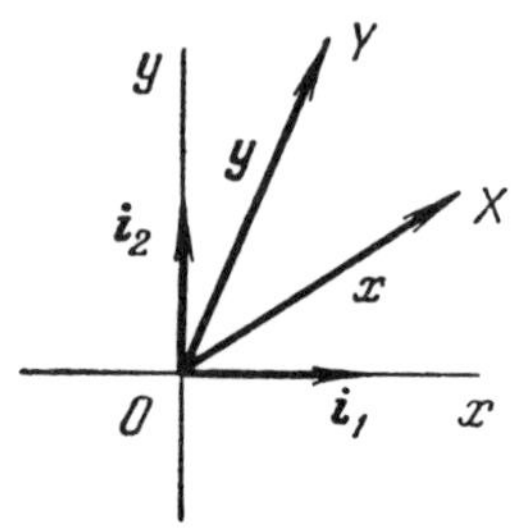

Figure 12.1.

from which it follows that

$$OX^2 + OY^2 - XY^2 = 2(x_1y_1 + x_2y_2). \tag{12.1}$$

Bearing in mind Pythagoras' theorem, we easily conclude on the basis of (12.1) that $\boldsymbol{x}$ and $\boldsymbol{y}$ are perpendicular if and only if

$$x_1y_1 + x_2y_2 = 0.$$

A similar argument applied to vectors in space yields the analogous perpendicularity condition

$$x_1y_1 + x_2y_2 + x_3y_3 = 0.$$

Formula (12.1) suggests that we associate with each pair of vectors $\boldsymbol{x}, \boldsymbol{y}$ in the plane the number

$$x_1y_1 + x_2y_2, \tag{12.2}$$

and in space the number

$$x_1y_1 + x_2y_2 + x_3y_3. \tag{12.2'}$$

In each case the number is called in geometry the *scalar product* of the vectors $\boldsymbol{x}$ and $\boldsymbol{y}$ and is denoted by $(\boldsymbol{x}, \boldsymbol{y})$.

We note that the length of a vector can be expressed by means of the scalar product. In fact, in the plane

$$|\boldsymbol{x}| = \sqrt{x_1^2 + x_2^2},$$

and in space

$$|\boldsymbol{x}| = \sqrt{x_1^2 + x_2^2 + x_3^2}.$$

In either case,

$$|\boldsymbol{x}| = \sqrt{(\boldsymbol{x}, \boldsymbol{x})}.$$

12.2 General Definition of the Scalar Product

Here are some of the simple properties of the scalar product of vectors in the plane and in space:

(1) $$(\boldsymbol{x}, \boldsymbol{x}) \geq 0. \ (\boldsymbol{x}, \boldsymbol{x}) = 0 \text{ only if } \boldsymbol{x} = \mathbf{o};$$

(2) $$(\boldsymbol{x}, \boldsymbol{y}) = (\boldsymbol{y}, \boldsymbol{x});$$

(3) $$(\boldsymbol{x}, k\boldsymbol{y}) = k(\boldsymbol{x}, \boldsymbol{y}), \text{ where k is any real number;}$$

(4) $$(\boldsymbol{x}, \boldsymbol{y} + \boldsymbol{z}) = (\boldsymbol{x}, \boldsymbol{y}) + (\boldsymbol{x}, \boldsymbol{z}).$$

The first three properties follow directly from the definition of the scalar product. The last property is not difficult to prove. Here is its proof for the case of vectors in space:

$$\begin{aligned} (\boldsymbol{x}, \boldsymbol{y} + \boldsymbol{z}) &= x_1(y_1 + z_1) + x_2(y_2 + z_2) + x_3(y_3 + z_3) \\ &= (x_1y_1 + x_2y_2 + x_3y_3) + (x_1z_1 + x_2z_2 + x_3z_3) \\ &= (\boldsymbol{x}, \boldsymbol{y}) + (\boldsymbol{x}, \boldsymbol{z}). \end{aligned}$$

We now come to the key issue of this section — the extension of the definition of the scalar product to the case of n dimensions. No matter how this is done it is desirable that properties (1) – (4) should hold. This guides the following definition.

Definition 12.1 *Suppose that with any two vectors $\boldsymbol{x}$ and $\boldsymbol{y}$ in the space* A_n *there is associated a number* $(\boldsymbol{x}, \boldsymbol{y})$ *such that the properties* (1), (2), (3), *and* (4) *hold. Then we say that a* scalar product *is given in* A_n *and call* $(\boldsymbol{x}, \boldsymbol{y})$ *the* scalar product of the vectors $\boldsymbol{x}$ and $\boldsymbol{y}$.

12.3 One Way of Introducing a Scalar Product

Our definition leaves open the question of the very possibility of introducing a scalar product in the space A_n. That this can be done, and how it can be done, is suggested by the expressions (12.2) and (12.2′). Specifically, let

$$\boldsymbol{a}_1, \boldsymbol{a}_2, \ldots, \boldsymbol{a}_n$$

be a basis in A_n. With any two vectors

$$\begin{aligned} \boldsymbol{x} &= x_1\boldsymbol{a}_1 + x_2\boldsymbol{a}_2 + \ldots + x_n\boldsymbol{a}_n, \\ \boldsymbol{y} &= y_1\boldsymbol{a}_1 + y_2\boldsymbol{a}_2 + \ldots + y_n\boldsymbol{a}_n, \end{aligned}$$

in A_n we associate the number

$$(\boldsymbol{x}, \boldsymbol{y}) = x_1y_1 + x_2y_2 + \ldots + x_ny_n. \tag{12.3}$$

It is not difficult to show that $(\boldsymbol{x}, \boldsymbol{y})$ satisfies the requirements (1)–(4), and is therefore a scalar product.

Let

$$\boldsymbol{a}'_1, \boldsymbol{a}'_2, \ldots, \boldsymbol{a}'_n$$

be another basis in A_n. Let the coordinates of $\boldsymbol{x}$ and $\boldsymbol{y}$ relative to this basis be $x'_1, x'_2, \ldots, x'_n$ and $y'_1, y'_2, \ldots, y'_n$, respectively. Then the equality

$$(\boldsymbol{x}, \boldsymbol{y})' = x'_1y'_1 + x'_2y'_2 + \ldots + x'_ny'_n$$

defines another scalar product in A_n. But, in general, it is *not* true that

$$(\boldsymbol{x}, \boldsymbol{y}) = (\boldsymbol{x}, \boldsymbol{y})'.$$

In other words, there are many scalar products in the space A_n. Nevertheless, as we shall show below, *the indicated manner of introducing a scalar product is general in the following sense: No matter how one introduces a scalar product in the space* A_n *there is a basis (in fact, there are many bases) in which formula* (12.3) *holds.*

12.4 Length of a Vector. Orthogonal Vectors

Given a scalar product, we define the length of a vector and the perpendicularity of two vectors by analogy with the two- and three-dimensional cases. Thus, by the *length*, or *norm*, of an n-dimensional vector we mean the number

$$|\boldsymbol{x}| = \sqrt{(\boldsymbol{x}, \boldsymbol{x})}$$

(note that, in view of property (1), the number under the square root sign is nonnegative), and we say of two vectors $\boldsymbol{x}$ and $\boldsymbol{y}$ that they are *perpendicular*, or *orthogonal* — in symbols, $\boldsymbol{x} \perp \boldsymbol{y}$ — if their scalar product is zero. In other words,

$$\boldsymbol{x} \perp \boldsymbol{y} \text{ means that } (\boldsymbol{x}, \boldsymbol{y}) = 0.$$

12.5 Expressing a Scalar Product in Terms of Coordinates

First we supplement properties (3) and (4) by the properties

$$(3') \qquad (k\boldsymbol{x}, \boldsymbol{y}) = k(\boldsymbol{x}, \boldsymbol{y});$$

$$(4') \qquad (\boldsymbol{x} + \boldsymbol{y}, \boldsymbol{z}) = (\boldsymbol{x}, \boldsymbol{z}) + (\boldsymbol{y}, \boldsymbol{z}).$$

Property (3′) follows from the chain of equalities

$$(k\boldsymbol{x}, \boldsymbol{y}) = (\boldsymbol{y}, k\boldsymbol{x}) = k(\boldsymbol{y}, \boldsymbol{x}) = k(\boldsymbol{x}, \boldsymbol{y}),$$

each of which is justified by some scalar product property. Similarly, (4′) follows from the chain of equalities

$$(\boldsymbol{x} + \boldsymbol{y}, \boldsymbol{z}) = (\boldsymbol{z}, \boldsymbol{x} + \boldsymbol{y}) = (\boldsymbol{z}, \boldsymbol{x}) + (\boldsymbol{z}, \boldsymbol{y}) = (\boldsymbol{x}, \boldsymbol{z}) + (\boldsymbol{y}, \boldsymbol{z}).$$

Combining properties (3) and (3′) we obtain

$$(3'') \qquad (k\boldsymbol{x}, l\boldsymbol{y}) = kl(\boldsymbol{x}, \boldsymbol{y}).$$

Further, (4) and (4′) imply that

$$(\boldsymbol{x}_1 + \boldsymbol{x}_2 + \ldots + \boldsymbol{x}_p, \ \boldsymbol{y}_1 + \boldsymbol{y}_2 + \ldots + \boldsymbol{y}_q) = \textstyle\sum_{i,j}(\boldsymbol{x}_i, \boldsymbol{y}_j),$$

that is, the scalar product of two sums is the sum of the scalar products of each of the summands of the first sum by each of the summands in the second sum. This and property (3″) justify the following rule for obtaining the scalar product of two linear combinations:

$$(k_1\boldsymbol{x}_1 + k_2\boldsymbol{x}_2 + \ldots + k_p\boldsymbol{x}_p, \ l_1\boldsymbol{y}_1 + l_2\boldsymbol{y}_2 + \ldots + l_q\boldsymbol{y}_q) = \\ = \sum_{i,j} k_i l_j (\boldsymbol{x}_i, \boldsymbol{y}_j). \qquad (12.4)$$

Now we can easily obtain an expression for the scalar product $(\boldsymbol{x}, \boldsymbol{y})$ in terms of the coordinates of $\boldsymbol{x}$ and $\boldsymbol{y}$. Specifically, let

$$\boldsymbol{a}_1, \boldsymbol{a}_2, \ldots, \boldsymbol{a}_n. \qquad (12.5)$$

be a basis of the space A_n and let

$$\begin{aligned} \boldsymbol{x} &= x_1\boldsymbol{a}_1 + x_2\boldsymbol{a}_2 + \ldots + x_n\boldsymbol{a}_n, \\ \boldsymbol{y} &= y_1\boldsymbol{a}_1 + y_2\boldsymbol{a}_2 + \ldots + y_n\boldsymbol{a}_n, \end{aligned}$$

be any two vectors in A_n. By formula (12.4)

$$(\boldsymbol{x}, \boldsymbol{y}) = \sum_{i,j} x_i y_j (\boldsymbol{a}_i, \boldsymbol{a}_j).$$

The quantities

$$g_{ij} = (\boldsymbol{a}_i, \boldsymbol{a}_j)$$

are constants depending on the choice of basis. Once a basis has been selected, the scalar product is given by the expression

$$(\boldsymbol{x}, \boldsymbol{y}) = \sum_{i,j} g_{ij} x_i y_j. \tag{12.6}$$

We shall apply this very useful result to prove many important propositions.

12.6 Existence of a Vector Orthogonal to p Given Vectors, $p < n$

We wish to find a vector $\boldsymbol{x}$ perpendicular to a given vector $\boldsymbol{y}$, that is, a vector such that $(\boldsymbol{x}, \boldsymbol{y}) = 0$. In view of (12.6), the coordinates $x_1, x_2, \ldots, x_n$ of the vector must satisfy the equation

$$\sum_{i,j} g_{ij} x_i y_j = 0.$$

Since the g_{ij} and the y_j are given numbers, the left side of our equation reduces to an expression of the form $a_1x_1 + a_2x_2 + \ldots + a_nx_n$. This means that our equation is a linear homogeneous equation in the variables $x_1, x_2, \ldots, x_n$.

If the vector $\boldsymbol{x}$ is to be orthogonal to p given vectors

$$\boldsymbol{y}_1, \boldsymbol{y}_2, \ldots, \boldsymbol{y}_p,$$

then its coordinates must satisfy a system of p linear homogeneous equations. By the lemma in chapter 11, such a system must have a nonzero solution provided that $p < n$. This implies the following theorem.

Theorem 12.1 *let*

$$\boldsymbol{y}_1, \boldsymbol{y}_2, \ldots, \boldsymbol{y}_p$$

be p given vectors in the space A_n. If $p < n$, then there exists a nonzero vector $\boldsymbol{x}$ perpendicular to all the given vectors.[3]

This theorem has many consequences. We shall consider just one of them that will play an important part in the sequel.

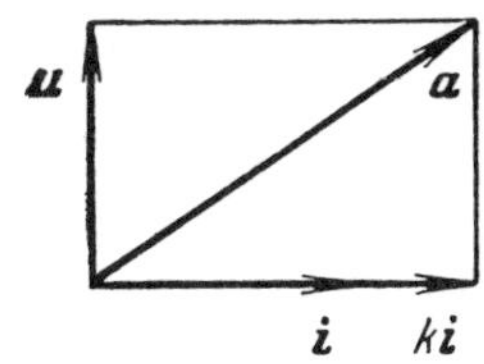

Figure 12.2.

Corollary 12.1 *If* P *is a subspace of the space* A_n *and* $P \neq A_n$*, then there exists a nonzero vector* $\boldsymbol{x} \in A_n$ *orthogonal to all vectors in* P. (We say, briefly, that $\boldsymbol{x}$ is orthogonal to the subspace P.)

The proof is almost obvious. Let

$$\boldsymbol{y}_1, \boldsymbol{y}_2, \ldots, \boldsymbol{y}_p$$

be a basis in P. *Since* $P \neq A_n$, we have $p < n$. But then there is a vector $\boldsymbol{x} \neq \mathbf{o}$ that is orthogonal to the vectors $\boldsymbol{y}_1, \boldsymbol{y}_2, \ldots, \boldsymbol{y}_p$. In turn, $\boldsymbol{x}$ is orthogonal to every linear combination of these vectors,

$$\begin{aligned}&(\boldsymbol{x}, k_1\boldsymbol{y}_1 + k_2\boldsymbol{y}_2 + \ldots + k_p\boldsymbol{y}_p)\\ &\quad = k_1(\boldsymbol{x}, \boldsymbol{y}_1) + k_2(\boldsymbol{x}, \boldsymbol{y}_2) + \ldots + k_p(\boldsymbol{x}, \boldsymbol{y}_p) = 0,\end{aligned}$$

and therefore to all of P.

12.7 Decomposition of a Vector into Two Components

We are about to prove in A_n a fact that is geometrically obvious in the plane as well as in space.

Let $\boldsymbol{i}$ be a nonzero vector.*Any vector* $\boldsymbol{a}$ *can be decomposed into a sum of two vectors of which one is a multiple of* $\boldsymbol{i}$ *and the other is perpendicular to* $\boldsymbol{i}$ (Figure 12.2.)

$$\boldsymbol{a} = k\boldsymbol{i} + \boldsymbol{u}, \qquad \boldsymbol{u} \perp \boldsymbol{i}.$$

To prove this assertion we must prove the existence of a number k such that the vector $\boldsymbol{u} = \boldsymbol{a} - k\boldsymbol{i}$ is orthogonal to $\boldsymbol{i}$, that is, such that

$$(\boldsymbol{a} - k\boldsymbol{i},\ \boldsymbol{i}) = 0.$$

Equivalently,

$$(\boldsymbol{a}, \boldsymbol{i}) = k(\boldsymbol{i}, \boldsymbol{i}).$$

But then

$$k = \frac{(\boldsymbol{a}, \boldsymbol{i})}{(\boldsymbol{i}, \boldsymbol{i})}.$$

(Note that $\boldsymbol{i} \neq \mathbf{o}$, so that $(\boldsymbol{i}, \boldsymbol{i}) \neq 0$.)

Chapter 13

Orthonormal Basis. Orthogonal Transformation

13.1 Definition of an Orthonormal Basis

We know that there are infinitely many bases in the space A_n. Before the introduction of the scalar product in A_n we had no reason for singling out any of them. After the introduction of the scalar product, however, the so-called orthonormal bases play a special role.

A basis

$$\boldsymbol{a}_1, \boldsymbol{a}_2, \ldots, \boldsymbol{a}_n$$

is said to be *orthonormal* if any two of its vectors are orthogonal,

$$(\boldsymbol{a}_i, \boldsymbol{a}_j) = 0 \quad (i, j = 1, \ldots, n; \; i \neq j), \tag{13.1}$$

and each of its vectors has length 1,

$$(\boldsymbol{a}_i, \boldsymbol{a}_i) = 1 \quad (i = 1, \ldots, n). \tag{13.2}$$

In ordinary 3-dimensional space an orthonormal basis consists of a triple of pairwise orthogonal unit vectors (Figure 13.1).

(The word "orthonormal" is composed of the words "orthogonal" and "normalized." A vector is said to be *normalized*, or a *unit* vector, if its length is 1.)

What makes an orthonormal basis special is the simplicity of the expression for the scalar product in such a basis. Specifically, if our basis

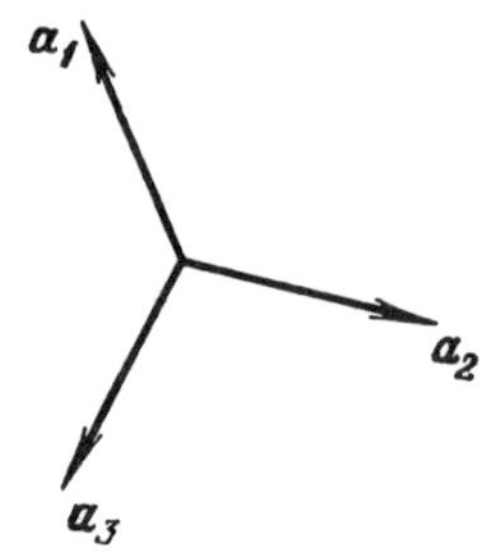

Figure 13.1.

is orthonormal, that is, if the equalities (13.1) and (13.2) hold, then the expression

$$(\boldsymbol{x}, \boldsymbol{y}) = \sum_{i,j} x_i y_j (\boldsymbol{a}_i, \boldsymbol{a}_j)$$

for the scalar product of two vectors reduces to

$$(\boldsymbol{x}, \boldsymbol{y}) = x_1 y_1 + x_2 y_2 + \ldots + x_n y_n; \tag{13.3}$$

that is, the scalar product of two vectors expressed in an orthonormal basis reduces to the sum of the products of the corresponding coordinates of the two vectors.

13.2 Existence of Orthonormal Bases

While we have established an important property of orthonormal bases we don't know whether such bases exist. We prove their existence next.

First a remark. Let $\boldsymbol{a}$ be a nonzero vector. Then the vector

$$\boldsymbol{a}' = \boldsymbol{a}/|\boldsymbol{a}|$$

has length one. Indeed,

$$(\boldsymbol{a}', \boldsymbol{a}') = (\boldsymbol{a}, \boldsymbol{a})/|\boldsymbol{a}|^2 = 1.$$

The transition from $\boldsymbol{a}$ to $\boldsymbol{a}'$ is called the *normalization* of the vector $\boldsymbol{a}$.

The existence of an orthonormal basis follows readily from a theorem proved in the previous chapter. Take a nonzero vector $\boldsymbol{b}_1$. Let $\boldsymbol{a}_1$ be the result of normalization of $\boldsymbol{b}_1$. Take a nonzero vector $\boldsymbol{b}_2$ orthogonal to

$\boldsymbol{a}_1$. Let $\boldsymbol{a}_2$ be the result of normalization of $\boldsymbol{b}_2$. Take a nonzero vector $\boldsymbol{b}_3$ orthogonal to $\boldsymbol{a}_1$ and $\boldsymbol{a}_2$. Let $\boldsymbol{a}_3$ be the result of the normalization of $\boldsymbol{b}_3$, and so on. Finally we obtain a vector $\boldsymbol{a}_n$ orthogonal to the vectors $\boldsymbol{a}_1, \boldsymbol{a}_2, \ldots, \boldsymbol{a}_{n-1}$. By theorem 9.2, the system of n vectors

$$\boldsymbol{a}_1, \boldsymbol{a}_2, \ldots, \boldsymbol{a}_n$$

is linearly independent. Suppose not. Then one of our vectors would be a linear combination of its predecessors. Specifically, suppose that

$$\boldsymbol{a}_3 = \alpha \boldsymbol{a}_1 + \beta \boldsymbol{a}_2.$$

Multiplying both sides of this equality by $\boldsymbol{a}_3$ we obtain

$$(\boldsymbol{a}_3, \boldsymbol{a}_3) = 0.$$

But this is impossible, for $(\boldsymbol{a}_3, \boldsymbol{a}_3) = 1$.

We note that by proving the existence of an orthonormal basis we have fulfilled a promise made in the previous chapter to demonstrate the *existence of a basis in which the scalar product is given by formula* (13.3).

13.3 A Method for Obtaining All Orthonormal Bases

▷ The study of orthonormal bases gives rise to a number of interesting questions. One of them is the question of the transition from one orthonormal basis to another. In chapter 9 we said that there is always a chain of elementary transformations that lead from one basis to another. This assertion applies, in particular, to orthonormal bases. But what spoils things here is that the result of applying an elementary transformation to an orthonormal basis is, in general, *not* an orthonormal basis. This can be remedied as follows.

Let $\boldsymbol{a}_1, \boldsymbol{a}_2, \ldots, \boldsymbol{a}_n$ be an orthonormal basis. By an *elementary orthotransformation* of a basis we mean

1. multiplication of a basis by (-1). (Clearly this transformation takes an orthonormal basis into an orthonormal basis);

2. the replacement of any two basis vectors $\boldsymbol{a}_i, \boldsymbol{a}_j (i \neq j)$ by vectors $\boldsymbol{a}'_i, \boldsymbol{a}'_j$ given by the formulas

$$\boldsymbol{a}'_i = \cos\alpha\, \boldsymbol{a}_i - \sin\alpha\, \boldsymbol{a}_j,$$

$$\boldsymbol{a}'_j = \sin\alpha\, \boldsymbol{a}_i - \cos\alpha\, \boldsymbol{a}_j,$$

where α is any real number.

We state without proof a theorem that explains the role of elementary orthotransformations.

Theorem 13.1 *Given any two orthogonal bases there is a chain of elementary orthotransformations leading from one to the other.* ◁

13.4 Orthogonal Transformations

The reader is probably familiar with the concept of a transformation (some transformations are studied, for example, in high school geometry courses). We are about to study certain transformations of the n-dimensional vector space A_n.

If a rule is given that associates with any vector $\boldsymbol{a} \in \mathrm{A}_n$ a vector $\boldsymbol{a}' \in \mathrm{A}_n$, then we say that there is given a *transformation of the space* A_n and write

$$\boldsymbol{a}' = F(\boldsymbol{a}).$$

Here F denotes the rule for obtaining $\boldsymbol{a}'$ from $\boldsymbol{a}$.

A transformation F is said to linear if it has the following two properties:

$$F(\boldsymbol{x} + \boldsymbol{y}) = F(\boldsymbol{x}) + F(\boldsymbol{y}), \tag{1}$$

$$F(k\boldsymbol{x}) = kF(\boldsymbol{x}). \tag{2}$$

Here $\boldsymbol{x}$ and $\boldsymbol{y}$ are any two vectors and k is any number. If we denote the vectors $F(\boldsymbol{x})$ and $F(\boldsymbol{y})$ by $\boldsymbol{x}'$ and $\boldsymbol{y}'$, then we can restate these conditions as follows: the transformation F maps the triple of vectors $\boldsymbol{x}$, $\boldsymbol{y}$, $\boldsymbol{x} + \boldsymbol{y}$ onto the triple $\boldsymbol{x}', \boldsymbol{y}', \boldsymbol{x}' + \boldsymbol{y}'$, and the pair $\boldsymbol{x}, k\boldsymbol{x}$ onto the pair $\boldsymbol{x}', k\boldsymbol{x}'$. In other words, a transformation is linear if it doesn't "disturb" either the sum of two vectors or the product of a vector by a number. It is clear that a linear transformation does not disturb linear combinations:

$$F(k_1\boldsymbol{x}_1 + k_2\boldsymbol{x}_2 + \ldots + k_p\boldsymbol{x}_p) \\ = k_1F(\boldsymbol{x}_1) + k_2F(\boldsymbol{x}_2) + \ldots + k_pF(\boldsymbol{x}_p).$$

Suppose that A_n is a space with a scalar product. Then the linear transformations of special interest are those that "preserve" the scalar product in the sense that

$$(F(\boldsymbol{x}), F(\boldsymbol{y})) = (\boldsymbol{x}, \boldsymbol{y}) \text{ for any } \boldsymbol{x}, \boldsymbol{y} \in \mathrm{A}_n.$$

Such transformations are called *orthogonal*.

It is clear that orthogonal transformations preserve the length of any vector $\boldsymbol{x}$, that is,

$$|F(\boldsymbol{x})| = |\boldsymbol{x}|.$$

This follows from the fact that the length of a vector is expressed in terms of the scalar product:

$$|\boldsymbol{x}| = \sqrt{(\boldsymbol{x}, \boldsymbol{x})}.$$

The property of preservation of the length of any vector may be taken as the defining property of an orthogonal transformation. This follows from the fact that the scalar product can be expressed in terms of lengths. In fact, the obvious identity

$$(\boldsymbol{x} + \boldsymbol{y}, \boldsymbol{x} + \boldsymbol{y}) = (\boldsymbol{x}, \boldsymbol{x}) + (\boldsymbol{y}, \boldsymbol{y}) + (\boldsymbol{x}, \boldsymbol{y}) + (\boldsymbol{y}, \boldsymbol{x})$$

implies that

$$\begin{aligned} 2(\boldsymbol{x}, \boldsymbol{y}) &= (\boldsymbol{x} + \boldsymbol{y}, \boldsymbol{x} + \boldsymbol{y}) - (\boldsymbol{x}, \boldsymbol{x}) - (\boldsymbol{y}, \boldsymbol{y}) \\ &= |\boldsymbol{x} + \boldsymbol{y}|^2 - |\boldsymbol{x}|^2 - |\boldsymbol{y}|^2. \end{aligned}$$

Orthogonal transformations have many important properties. We mention one of them.

Let $\boldsymbol{a}_1, \boldsymbol{a}_2, \dots, \boldsymbol{a}_n$ be an orthonormal basis and F an orthogonal transformation of A_n. Then the vectors

$$\boldsymbol{a}'_1 = F(\boldsymbol{a}_1), \quad \boldsymbol{a}'_2 = F(\boldsymbol{a}_2), \quad \dots, \quad \boldsymbol{a}'_n = F(\boldsymbol{a}_n)$$

also form an orthonormal basis of A_n. In other words, *orthogonal transformations map orthonormal bases onto orthonormal bases.*

Indeed, the orthogonality of the transformation F implies that

$$(\boldsymbol{a}_i, \boldsymbol{a}_j) = (\boldsymbol{a}'_i, \boldsymbol{a}'_j)$$

for all i, j from 1 to n. This means that the vectors $\boldsymbol{a}'_1, \boldsymbol{a}'_2, \dots, \boldsymbol{a}'_n$ satisfy the relations

$$(\boldsymbol{a}'_i, \boldsymbol{a}'_i) = 1, \quad (\boldsymbol{a}'_i, \boldsymbol{a}'_j) = 0 \qquad (i \neq j).$$

But then they also form an orthonormal basis (see the argument at the end of section 2).

13.5 The Inverse of an Orthogonal Transformation

First we establish the fact that if F is an orthogonal transformation and $\boldsymbol{b}$ is any vector in A_n, then the equation

$$F(\boldsymbol{x}) = \boldsymbol{b} \tag{13.4}$$

has a unique solution.

This can be proved as follows. Let $\boldsymbol{a}_1, \boldsymbol{a}_2, \ldots, \boldsymbol{a}_n$ be an orthonormal basis. Then the vectors $\boldsymbol{a}'_1 = F(\boldsymbol{a}_1)$, $\boldsymbol{a}'_2 = F(\boldsymbol{a}_2)$, $\ldots$, $\boldsymbol{a}'_n = F(\boldsymbol{a}_n)$, also form an orthonormal basis. If the vector $\boldsymbol{b}$ is given by

$$\boldsymbol{b} = k_1\boldsymbol{a}'_1 + k_2\boldsymbol{a}'_2 + \ldots + k_n\boldsymbol{a}'_n,$$

then the vector

$$\boldsymbol{a} = k_1\boldsymbol{a}_1 + k_2\boldsymbol{a}_2 + \ldots + k_n\boldsymbol{a}_n$$

is a solution of equation (13.4). Indeed, the linearity of F implies that

$$F(\boldsymbol{a}) = k_1\boldsymbol{a}'_1 + k_2\boldsymbol{a}'_2 + \ldots + k_n\boldsymbol{a}'_n = \boldsymbol{b}.$$

This proves that equation (13.4) has a solution. Its uniqueness is easy to establish, for if

$$F(\boldsymbol{x}_1) = \boldsymbol{b} \text{ and } F(\boldsymbol{x}_2) = \boldsymbol{b},$$

then $F(\boldsymbol{x}_1) = F(\boldsymbol{x}_2)$, so that, $F(\boldsymbol{x}_1 - \boldsymbol{x}_2) = \boldsymbol{o}$. But then $|\boldsymbol{x}_1 - \boldsymbol{x}_2| = 0$, that is, $\boldsymbol{x}_1 = \boldsymbol{x}_2$.

If with each vector $\boldsymbol{b} \in \mathrm{A}_n$ we associate the vector $\boldsymbol{x}$ that is the solution of equation (13.4), then we obtain a new transformation F^{-1} called the *inverse* of F. Another way of putting this is that *every orthogonal transformation has an inverse.*

It is natural to ask if the inverse of an orthogonal transformation is itself orthogonal. We shall prove that this is indeed the case.

The linearity of F^{-1} is an easy consequence of the linearity of F: if F maps the triple of vectors $\boldsymbol{x}, \boldsymbol{y}, \boldsymbol{x} + \boldsymbol{y}$ onto the triple $\boldsymbol{x}', \boldsymbol{y}', \boldsymbol{x}' + \boldsymbol{y}'$, then F^{-1} maps the triple $\boldsymbol{x}', \boldsymbol{y}', \boldsymbol{x}' + \boldsymbol{y}'$ onto the triple $\boldsymbol{x}, \boldsymbol{y}, \boldsymbol{x} + \boldsymbol{y}$, and if F maps the pair $\boldsymbol{x}, k\boldsymbol{x}$ onto the pair $\boldsymbol{x}', k\boldsymbol{x}'$, then F^{-1} maps the pair $\boldsymbol{x}', k\boldsymbol{x}'$ onto the pair $\boldsymbol{x}, k\boldsymbol{x}$. To see that F^{-1} preserves scalar products note that the equality

$$(\boldsymbol{x}, \boldsymbol{y}) = (\boldsymbol{x}', \boldsymbol{y}')$$

implies the equality

$$(F^{-1}(\boldsymbol{x}'), F^{-1}(\boldsymbol{y}')) = (\boldsymbol{x}', \boldsymbol{y}').$$

Since the transformation F^{-1} is linear and preserves scalar products, it is orthogonal.

We see that *the inverse of an orthogonal transformation is orthogonal.*

13.6 "How Many" Different Orthogonal Transformations Are There?

▷ It is natural to ask if there are orthogonal transformations and how large is the class of such transformations. The following proposition will help us answer these questions.

Let

$$\boldsymbol{a}_1, \boldsymbol{a}_2, \ldots, \boldsymbol{a}_n \text{ and } \boldsymbol{a}'_1, \boldsymbol{a}'_2, \ldots, \boldsymbol{a}'_n$$

be two orthonormal bases. Then there exists a unique orthogonal transformation that maps the first of these bases onto the second.

We define the required transformation as the transformation F that associates with any vector

$$\boldsymbol{a} = k_1\boldsymbol{a}_1 + k_2\boldsymbol{a}_2 + \ldots + k_n\boldsymbol{a}_n$$

the vector

$$F(\boldsymbol{a}) = k_1\boldsymbol{a}'_1 + k_2\boldsymbol{a}'_2 + \ldots + k_n\boldsymbol{a}'_n$$

and show that it is orthogonal.

The linearity of F is obvious. It remains to show that F preserves scalar products. Let

$$\boldsymbol{x} = x_1\boldsymbol{a}_1 + \ldots + x_n\boldsymbol{a}_n \text{ and } \boldsymbol{y} = y_1\boldsymbol{a}_1 + \ldots + y_n\boldsymbol{a}_n \tag{13.5}$$

be any two vectors in A_n. Then

$$F(\boldsymbol{x}) = x_1\boldsymbol{a}'_1 + \ldots + x_n\boldsymbol{a}'_n, \quad F(\boldsymbol{y}) = y_1\boldsymbol{a}'_1 + \ldots + y_n\boldsymbol{a}'_n. \tag{13.6}$$

Since the basis $\boldsymbol{a}_1, \boldsymbol{a}_2, \ldots, \boldsymbol{a}_n$ is orthonormal, (13.5) implies that

$$(\boldsymbol{x}, \boldsymbol{y}) = x_1y_1 + x_2y_2 + \ldots + x_ny_n.$$

Since the basis $\boldsymbol{a}'_1, \boldsymbol{a}'_2, \ldots, \boldsymbol{a}'_n$ is also orthonormal, (13.6) implies that

$$(F(\boldsymbol{x}), F(\boldsymbol{y})) = x_1y_1 + x_2y_2 + \ldots + x_ny_n.$$

But then

$$(\boldsymbol{x}, \boldsymbol{y}) = (F(\boldsymbol{x}), F(\boldsymbol{y})),$$

that is, the transformation F is orthogonal.

That there is just one orthogonal transformation that maps the first basis onto the second follows from the linearity of F. Specifically, the effect of a linear transformation on any vector is uniquely determined by its effect on the vectors of a basis.

This proposition justifies the following conclusion: it is possible to establish a one-to-one correspondence between the class of orthonormal bases and the class of orthogonal transformations. To do this fix some orthonormal basis B_o and associate to any orthonormal basis B the (unique) orthogonal transformation that maps B_o onto B. This shows that there are "as many" orthogonal transformation as there are orthonormal bases. $\triangleleft$

Part III

The Exceptional Position of Four Algebras

Certain algebras occupy a special position in the infinitude of algebras. They are the algebras $\mathcal{C}, \mathcal{Q}$ and $\mathcal{O}$ of complex numbers, quaternions, and Cayley numbers. There are many ways of describing the distinguishing features of these algebras but they all come down to the following: compared with other algebras, these three algebras are closest to what is, in a sense, their original foundation — the algebra $\mathcal{R}$ of real numbers. Examples of this closeness are:

1. The algebras $\mathcal{R}, \mathcal{C}$, and $\mathcal{Q}$ are the only division algebras with associative multiplication (briefly, $\mathcal{R}, \mathcal{C}$, and $\mathcal{Q}$ are the only associative division algebras). A somewhat more precise version of this proposition is known as Frobenius' theorem.

2. The algebras $\mathcal{R}, \mathcal{C}, \mathcal{Q}$, and $\mathcal{O}$ are the only division algebras in which the following formulas hold : $(\boldsymbol{uv})\boldsymbol{v} = \boldsymbol{u}(\boldsymbol{vv})$ and $\boldsymbol{v}(\boldsymbol{vu}) = (\boldsymbol{vv})\boldsymbol{u}$ (briefly, $\mathcal{R}, \mathcal{C}, \mathcal{Q}$, and $\mathcal{O}$ are the only alternative division algebras). This proposition is known as the generalized Frobenius theorem.

3. The algebras $\mathcal{R}, \mathcal{C}, \mathcal{Q}$, and $\mathcal{O}$ are the only algebras with an identity in which it is possible to define a scalar product such that the norm of a product is the product of the norms of the factors.

This is the substance of Hurwitz's theorem.

To put it differently, there is a certain hierarchy of algebras. Its very foundation is the algebra of real numbers. Its closest neighbor is the algebra of complex numbers in which multiplication retains the most important properties of the multiplication of real numbers such as commutativity, associativity, invertibility (this is an allusion to the possibility of division), and the existence of a multiplicative identity. Then comes the algebra of quaternions, in which multiplication is no longer commutative. Then comes the algebra of Cayley numbers, in which the multiplication is "alternative" rather than associative, but which is still a division algebra with a multiplicative identity. Other algebras do not enjoy such a "minimal package" of properties. Of course, this does not make them less interesting or important. It is simply that we happen to be concerned with what may be called the proximity of an algebra to the algebra of real numbers.

A final remark. In part 1 we formulated the "problem of the sum of squares," which consists in finding all identities of the form

$$\begin{aligned}(a_1^2 + a_2^2 + \ldots + a_n^2)&(b_1^2 + b_2^2 + \ldots + b_n^2) \\ &= \Phi_1^2 + \Phi_2^2 + \ldots + \Phi_n^2 \qquad (13.7)\end{aligned}$$

(see chapter 3). Starting with the "norm property" (the norm of a product is the product of the norms of the factors) of the algebras $\mathcal{R}, \mathcal{C}, \mathcal{Q}$, and $\mathcal{O}$, we constructed in part 1 concrete examples of such identities for $n = 1, 2, 4, 8$. In the present part we shall show that the number n in the identity (13.7) can take on just these four values. The proof of this fact is very much a consequence of Hurwitz's theorem, so that the "heroes" of the problem of the sum of squares are once more the algebras $\mathcal{R}, \mathcal{C}, \mathcal{Q}$, and $\mathcal{O}$.

In this part we make precise and prove all the facts mentioned above.

Chapter 14

Isomorphic Algebras

According to the definition in chapter 7, any n-dimensional algebra consists of elements that are uniquely representable in the form

$$a_1\boldsymbol{i}_1 + a_2\boldsymbol{i}_2 + \ldots + a_n\boldsymbol{i}_n$$

and are added and multiplied by real numbers according to natural rules. In other words, an n-dimensional algebra is, first of all, an n-dimensional vector space. Beyond that, there is given a multiplication table of the (initial) basis elements $\boldsymbol{i}_1, \boldsymbol{i}_2, \ldots, \boldsymbol{i}_n$, that is, a table of n^2 relations

$$\boldsymbol{i}_\alpha\boldsymbol{i}_\beta = k_{\alpha\beta,1}\boldsymbol{i}_1 + k_{\alpha\beta,2}\boldsymbol{i}_2 + \ldots + k_{\alpha\beta,n}\boldsymbol{i}_n \qquad (\alpha, \beta = 1, 2, \ldots, n), \tag{14.1}$$

where $k_{\alpha\beta,\gamma}$ are certain real numbers. Given the rules of multiplication of the basis elements, we multiply any two elements

$$a_1\boldsymbol{i}_1 + \ldots + a_n\boldsymbol{i}_n \text{ and } b_1\boldsymbol{i}_1 + \ldots + b_n\boldsymbol{i}_n$$

of the algebra by following the usual rule of multiplication of sums and then taking into consideration the relations (14.1).

In sum, we can say that an n-dimensional algebra is an n-dimensional vector space with a multiplication table (14.1) of the basis elements.

It would seem that two n-dimensional algebras with *different* multiplication tables should be regarded as different algebras. But this would not be entirely appropriate for reasons that follow.

Consider an n-dimensional algebra with initial basis $\boldsymbol{i}_1, \boldsymbol{i}_2, \ldots, \boldsymbol{i}_n$ and multiplication table (14.1). If we select in $\mathcal{A}$ another basis $\boldsymbol{i}'_1, \boldsymbol{i}'_2, \ldots, \boldsymbol{i}'_n$

then, of course, we shall have some other multiplication table

$$\boldsymbol{i}'_\alpha \boldsymbol{i}'_\beta = l_{\alpha\beta,1}\boldsymbol{i}'_1 + l_{\alpha\beta,2}\boldsymbol{i}'_2 + \ldots + l_{\alpha\beta,n}\boldsymbol{i}'_n \qquad (\alpha, \beta = 1, 2, \ldots, n). \tag{14.2}$$

Now consider some algebra $\mathcal{A}'$ with initial basis $\boldsymbol{i}'_1, \boldsymbol{i}'_2, \ldots, \boldsymbol{i}'_n$ and multiplication table (14.2). Should we regard $\mathcal{A}'$ as different from $\mathcal{A}$? In purely formal terms the answer is obviously yes. On the other hand, $\mathcal{A}'$ is essentially just the algebra $\mathcal{A}$ referred to a different basis, and there is every reason to regard the difference between the two algebras as inessential. This point of view is reflected in the concept of an isomorphism.

Definition 14.1 *Two n-dimensional algebras are said to be isomorphic if they have bases with identical multiplication tables.*

Of course, the sameness of the multiplication tables need not imply the same designations for the corresponding basis elements; the basis elements of one algebra may be denoted as $\boldsymbol{c}_1, \boldsymbol{c}_2, \ldots, \boldsymbol{c}_n$ and those of the other as $\boldsymbol{d}_1, \boldsymbol{d}_2, \ldots, \boldsymbol{d}_n$. But the coefficients of the linear combination of the $\boldsymbol{c}$'s representing a product $\boldsymbol{c}_\alpha \boldsymbol{c}_\beta$ must be the same as the corresponding coefficients of the $\boldsymbol{d}$'s representing the product $\boldsymbol{d}_\alpha \boldsymbol{d}_\beta$.

For example, if

$$\boldsymbol{c}_3 \boldsymbol{c}_2 = 3\boldsymbol{c}_1 - 7\boldsymbol{c}_5$$

then

$$\boldsymbol{d}_3 \boldsymbol{d}_2 = 3\boldsymbol{d}_1 - 7\boldsymbol{d}_5.$$

In mathematics *two isomorphic algebras are not regarded as different. Rather, they are thought of as two different copies of the same algebra.* This means that the answer to the problem of finding all algebras with a certain special property must have the form: an algebra with the required property is isomorphic to one concrete algebra, or another, or a third, and so on.

In part 1 we introduced the concept of a hypercomplex system and then the broader concept of an algebra. Now that we have at our disposal the concept of an isomorphism, we can make the relation between these concepts completely clear. In chapter 7 we showed that every hypercomplex system may be viewed as an algebra in which the first initial basis element is replaced by the identity element of the algebra, that is,

$$\boldsymbol{i}_1 \boldsymbol{i}_\alpha = \boldsymbol{i}_\alpha \boldsymbol{i}_1 = \boldsymbol{i}_\alpha$$

for all α. Now we can supplement this with a kind of converse: Every algebra with an identity is isomorphic to some hypercomplex system. In fact, given an algebra with an identity $\mathbf{1}$ we can choose in it a basis

$\boldsymbol{i}'_1, \boldsymbol{i}'_2, \ldots, \boldsymbol{i}'_n$ with $\boldsymbol{i}'_1 = \mathbf{1}$. Then we obtain a multiplication table in which $\boldsymbol{i}'_1 \boldsymbol{i}'_\alpha = \boldsymbol{i}'_\alpha \boldsymbol{i}'_1 = \boldsymbol{i}'_\alpha$ for all α, that is, the multiplication table of some hypercomplex system $\mathcal{M}$. It follows that the initial algebra is isomorphic to the hypercomplex system $\mathcal{M}$.

We conclude with an example that illustrates the role of the concept of isomorphism. We can now describe the result of chapter 2 by saying that every 2-dimensional algebra with an identity is isomorphic to one of the three algebras of complex, double, or dual numbers. This is the precise rendering of the statement in section 2.2 to the effect that "every system of numbers $a + bi$ with the operation rules 1 to 3 reduces to one of the following three ..." "Reduces to" is now replaced by "is isomorphic to."

Chapter 15

Subalgebras

In part 1 we have more than once encountered the phenomenon of one algebra being part of another. For example, the algebra of real numbers is part of the algebra of complex numbers, which is part of the algebra of quaternions, which itself is part of the algebra of Cayley numbers, and so on. In such cases we use the term "subalgebra" instead of "part."

Definition 15.1 *A set $\mathcal{P}$ of elements of an algebra $\mathcal{A}$ is called* subalgebra *of $\mathcal{A}$ if*

1. *$\mathcal{P}$ is a subspace of the vector space $\mathcal{A}$;*
2. *$\mathcal{P}$ is closed under the multiplication in $\mathcal{A}$, that is, if $\boldsymbol{a} \in P$ and $\boldsymbol{b} \in \mathcal{P}$, then $\boldsymbol{ab} \in \mathcal{P}$.*

The first requirement is equivalent (see chapter 10) to the condition that $\mathcal{P}$ is the totality of linear combinations

$$k_1\boldsymbol{a}_1 + k_2\boldsymbol{a}_2 + \ldots + k_p\boldsymbol{a}_p$$

of some elements $\boldsymbol{a}_1, \boldsymbol{a}_2, \ldots, \boldsymbol{a}_p$. The latter may be taken to be linearly independent. Then they form a basis of the subspace $\mathcal{P}$ (and their number does not exceed n).

To satisfy the second condition it suffices that all products

$$\boldsymbol{a}_\alpha \boldsymbol{a}_\beta \quad (\alpha, \beta = 1, 2, \ldots, p)$$

of the basis elements are again in $\mathcal{P}$, that is, that

$$\boldsymbol{a}_\alpha \boldsymbol{a}_\beta = k_{\alpha\beta,1}\boldsymbol{a}_1 + k_{\alpha\beta,2}\boldsymbol{a}_2 + k_{\alpha\beta,p}\boldsymbol{a}_p \quad (\alpha, \beta = 1, \ldots, p). \tag{15.1}$$

Our definition implies that a subalgebra may be regarded as an algebra in its own right with initial basis $\boldsymbol{a}_1, \boldsymbol{a}_2, \ldots, \boldsymbol{a}_p$ and multiplication table (15.1).

We give examples of subalgebras.

1. In the algebra of quaternions the subspace with basis $\mathbf{1}, \boldsymbol{j}$ is a subalgebra. More generally, any subspace with basis $\mathbf{1}, \boldsymbol{q}$, where $\boldsymbol{q}$ is not a multiple of $\mathbf{1}$, is a subalgebra. Each of these subalgebras is isomorphic to the algebra of complex numbers.

2. In the algebra of Cayley numbers the subspace with basis $\mathbf{1}, \boldsymbol{i}, \boldsymbol{E}, \boldsymbol{I}$ is a subalgebra. This subalgebra is isomorphic to the algebra of quaternions (the multiplication of the elements of this basis is the same as the multiplication table of the elements of the quaternion basis $\mathbf{1}, \boldsymbol{i}, \boldsymbol{j}, \boldsymbol{k}$). Similar examples are furnished by spaces with bases $\mathbf{1}, \boldsymbol{a}, \boldsymbol{b}, \boldsymbol{ab}$, where $\boldsymbol{a}$ and $\boldsymbol{b}$ are any two imaginary units from the initial basis $\mathbf{1}, \boldsymbol{i}, \boldsymbol{j}, \boldsymbol{k}, \boldsymbol{E}, \boldsymbol{I}, \boldsymbol{J}, \boldsymbol{K}$ of the algebra of Cayley numbers.

3. In the algebra of matrices of order n the matrices with zero in the first k rows, k fixed, from a subalgebra. A more complicated example is furnished by the subspace of all "chessboard" matrices, that is, matrices in which the elements a_{ij} with $i+j$ an odd integer are zero. For example, for $n=3$ these are the matrices of the form

$$\begin{pmatrix} * & 0 & * \\ 0 & * & 0 \\ * & 0 & * \end{pmatrix}.$$

We leave the verification of this fact to the reader.

Chapter 16

Translation of the "Problem of the Sum of Squares" into the Language of Algebras. Normed Algebras

We recall the formulation of the problem of the sum of squares posed in part 1. It is required to find out for what values of n and for what n bilinear forms

$$\begin{array}{l}\Phi_1(x_1, x_2, \dots, x_n;\ \ y_1, y_2, \dots, y_n),\\ \Phi_2(x_1, x_2, \dots, x_n;\ \ y_1, y_2, \dots, y_n),\\ \dots\dots\dots\dots\dots\dots\dots\dots\dots\dots\\ \Phi_n(x_1, x_2, \dots, x_n;\ \ y_1, y_2, \dots, y_n),\end{array}$$

we have the identity

$$(x_1^2 + x_2^2 + \dots + x_n^2)(y_1^2 + y_2^2 + \dots + y_n^2) = \Phi_1^2 + \Phi_2^2 + \dots + \Phi_n^2 (!)$$

In part 1, the study of certain concrete algebras (the algebras of complex numbers, quaternions, and Cayley numbers) enabled us to construct examples of the identity (!) for $n = 2, 4, 8$. But we said nothing about the construction of an *arbitrary* identity (!). We consider this issue next.

16.1 The Connection between (!) and a Certain Algebra $\mathcal{A}$

First we note that with every identity (!) there is associated a certain algebra defined in the following manner. We consider the n-dimensional vector space whose elements are the vectors

$$x_1 \boldsymbol{i}_1 + x_2 \boldsymbol{i}_2 + \ldots + x_n \boldsymbol{i}_n. \tag{16.1}$$

The product of two elements

$$\boldsymbol{x} = x_1 \boldsymbol{i}_1 + x_2 \boldsymbol{i}_2 + \ldots + x_n \boldsymbol{i}_n$$

and

$$\boldsymbol{y} = y_1 \boldsymbol{i}_1 + y_2 \boldsymbol{i}_2 + \ldots + y_n \boldsymbol{i}_n$$

in that space is defined by the formula

$$\boldsymbol{xy} = \Phi_1 \boldsymbol{i}_1 + \Phi_2 \boldsymbol{i}_2 + \ldots + \Phi_n \boldsymbol{i}_n. \tag{16.2}$$

In view of the linearity of the forms $\Phi_1, \Phi_2, \ldots, \Phi_n$ with respect to the variables $x_1, x_2, \ldots, x_n$ as well as the variables $y_1, y_2, \ldots, y_n$ it is clear that the following equalities hold:

$$k\boldsymbol{x} \cdot \boldsymbol{y} = k(\boldsymbol{xy}) \qquad \boldsymbol{x} \cdot k\boldsymbol{y} = k(\boldsymbol{xy}),$$
$$(\boldsymbol{x}_1 + \boldsymbol{x}_2)\boldsymbol{y} = \boldsymbol{x}_1\boldsymbol{y} + \boldsymbol{x}_2\boldsymbol{y}, \quad \boldsymbol{x}(\boldsymbol{y}_1 + \boldsymbol{y}_2) = \boldsymbol{x}\boldsymbol{y}_1 + \boldsymbol{x}\boldsymbol{y}_2.$$

But then the multiplication rule (16.2) actually defines a certain algebra (see section 7.7). Let this algebra be denoted by $\mathcal{A}$. From what we said above it follows that the algebra $\mathcal{A}$ is completely determined by the identity (!).

16.2 The Possibility of Introducing a Norm in the Algebra $\mathcal{A}$

We wish to find out what property of the algebra $\mathcal{A}$ is a reflection of the fact that forms $\Phi_1, \Phi_2, \ldots, \Phi_n$ are not entirely arbitrary but satisfy the identity (!).

To this end we introduce in the algebra $\mathcal{A}$ a scalar product $(\boldsymbol{x}, \boldsymbol{y})$ defined in terms of the coordinates of the vectors $\boldsymbol{x}$ and $\boldsymbol{y}$ relative to the basis $\boldsymbol{i}_1, \boldsymbol{i}_2, \ldots, \boldsymbol{i}_n$ by means of the rule

$$(\boldsymbol{x}, \boldsymbol{y}) = x_1 y_1 + x_2 y_2 + \ldots + x_n y_n. \tag{16.3}$$

In particular,

$$(\boldsymbol{x}, \boldsymbol{x}) = x_1^2 + x_2^2 + \ldots + x_n^2.$$

We note that by defining the scalar product in this way we make the basis $\boldsymbol{i}_1, \boldsymbol{i}_2, \ldots, \boldsymbol{i}_n$ orthonormal. Indeed,

$$(\boldsymbol{i}_\alpha, \boldsymbol{i}_\alpha) = 1,$$
$$(\boldsymbol{i}_\alpha, \boldsymbol{i}_\beta) = 0,$$

for $\alpha, \beta = 1, \ldots, n, \quad \alpha \neq \beta$. This is so because the only nonzero coordinate of the vector $\boldsymbol{i}_\alpha$ is its $\alpha - th$ coordinate (it has the value 1), and the only nonzero coordinate of $\boldsymbol{i}_\beta$ is its $\beta - th$ coordinate.

Now that we have introduced in the algebra $\mathcal{A}$ a scalar product, we can give the identity (!) a new interpretation. It is easy to see that the expression on the right side of the identity is the "scalar-product square" $(\boldsymbol{xy}, \boldsymbol{xy})$ of the element $\boldsymbol{xy}$, and the left side is the product of the scalar-product squares $(\boldsymbol{x}, \boldsymbol{x})$ and $(\boldsymbol{y}, \boldsymbol{y})$. This means that we can write (!) as

$$(\boldsymbol{xy}, \boldsymbol{xy}) = (\boldsymbol{x}, \boldsymbol{x})(\boldsymbol{y}, \boldsymbol{y}). \tag{16.4}$$

By defining the norm of an element $\boldsymbol{x}$ by the formula

$$|\boldsymbol{x}| = \sqrt{(\boldsymbol{x}, \boldsymbol{x})},$$

we can rewrite (16.4) as

$$|\boldsymbol{xy}| = |\boldsymbol{x}||\boldsymbol{y}| \tag{16.4'}$$

(the norm of a product is the product of the norms of the factors).

Next we make the following

Definition 16.1 *We say that an algebra $\mathcal{A}$ is* normed *if we can define in it a scalar product such that the identity* (16.4) *holds.*

Examples of normed algebras are the by now familiar algebras of complex numbers, quaternions, and Cayley numbers. That these are normed algebras follows from the fact that formula (16.4′) holds in them.

In order to satisfy all the requirements of the definition of a normed algebra we need only introduce a scalar product such that $|\boldsymbol{x}| = \sqrt{(\boldsymbol{x}, \boldsymbol{x})}$. For complex numbers such a scalar product is given by the formula

$$(\boldsymbol{z}, \boldsymbol{z}') = x_1 y_1 + x_2 y_2,$$

where $\boldsymbol{z} = x_1 + y_1 \boldsymbol{i}, \quad \boldsymbol{z}' = x_2 + y_2 \boldsymbol{i}$, and for the algebra of quaternions it is given by the formula

$$(\boldsymbol{q}, \boldsymbol{q}') = x_1 y_1 + x_2 y_2 + x_3 y_3 + x_4 y_4,$$

where $\boldsymbol{q} = x_1 + x_2 \boldsymbol{i} + x_3 \boldsymbol{j} + x_4 \boldsymbol{k}, \quad \boldsymbol{q}' = y_1 + y_2 \boldsymbol{i} + y_3 \boldsymbol{j} + y_4 \boldsymbol{k}$. There is an analogous definition of a scalar product for the algebra of Cayley numbers.

16.3 Conclusion

We have shown that to every identity (!) there corresponds a certain normed algebra $\mathcal{A}$. In that algebra the product of two elements $\boldsymbol{x} = x_1\boldsymbol{i}_1 + \ldots + x_n\boldsymbol{i}_n$, $\boldsymbol{y} = y_1\boldsymbol{i}_1 + \ldots + y_n\boldsymbol{i}_n$ is defined by the formula

$$\boldsymbol{xy} = \Phi_1\boldsymbol{i}_1 + \ldots + \Phi_n\boldsymbol{i}_n, \tag{16.5}$$

and the scalar product by the formula

$$(\boldsymbol{x}, \boldsymbol{y}) = x_1y_1 + x_2y_2 + \ldots + x_ny_n.$$

In the algebra $\mathcal{A}$ the elements $\boldsymbol{i}_1, \boldsymbol{i}_2, \ldots, \boldsymbol{i}_n$ form an orthonormal basis. Also, the identity (!) expresses the condition of normability relative to that basis.

It is easy to see that the converse of this proposition holds. Specifically, let us select in a given normed algebra $\mathcal{A}$ an orthonormal basis $\boldsymbol{i}_1, \boldsymbol{i}_2, \ldots, \boldsymbol{i}_n$. If we write down the rule of multiplication relative to this basis, then we obtain n forms $\Phi_1, \Phi_2, \ldots, \Phi_n$, and if we write down the normability condition for the algebra $\mathcal{A}$, then we obtain the identity (!) with the forms $\Phi_1, \Phi_2, \ldots, \Phi_n$ on the right side.

In sum, we arrive at the following conclusion.

All n-tuples of forms $\Phi_1, \Phi_2, \ldots, \Phi_n$ satisfying the identity (!) *can be obtained in the following manner: We take any normed n-dimensional algebra $\mathcal{A}$ and choose in it an orthonormal basis $\boldsymbol{i}_1, \boldsymbol{i}_2, \ldots, \boldsymbol{i}_n$. Then we write down the law of multiplication in the algebra $\mathcal{A}$ in the form* (16.5).

It follows that the problem of determining all identities (!) reduces to two problems:

1. finding all normed algebras.

2. writing down the multiplication law for each of these algebras relative to all orthonormal bases.

We shall consider the first of these problems in the next two chapters. We shall use the solution of the first problem to obtain a survey of all identities (!).

Chapter 17

Normed Algebras with an Identity. Hurwitz's Theorem

17.1 Formulation of Hurwitz's Theorem

In the previous chapter, in discussing the "problem of the sum of squares," we concluded that it was necessary to find all normed algebras. In this connection we prove a theorem, first established by the German mathematician A. Hurwitz in 1896, which does not give us a complete survey of normed algebras but disposes of a large part of the difficulties associated with this problem.

Hurwitz's theorem. *Every normed algebra with an identity is isomorphic to one of following four algebras: the real numbers, the complex numbers, the quaternions, and the Cayley numbers.*

The requirement that the algebra has an identity is essential and cannot be left out. We shall see later that there exist algebras without identities, and none of them can be isomorphic to one of the algebras mentioned in the theorem, all of which have identities.

The proof of Hurwitz's theorem is quite long. That is why we first present the overall scheme showing the ideas of the proof and fill in the details later.

17.2 Sketch of the Proof of Hurwitz's Theorem

Let $\mathcal{A}$ be a normed algebra with an identity. We recall that a normed algebra is an algebra in which one can define a scalar product such that

$$(\boldsymbol{ab}, \boldsymbol{ab}) = (\boldsymbol{a}, \boldsymbol{a})(\boldsymbol{b}, \boldsymbol{b}). \tag{17.1}$$

Let $\mathbf{1}$ be the identity of the algebra $\mathcal{A}$. Every element $\boldsymbol{a} \in \mathcal{A}$ can be uniquely represented as a sum of two terms[4] one of which is proportional to $\mathbf{1}$ and the other orthogonal to $\mathbf{1}$. Thus

$$\boldsymbol{a} = k\mathbf{1} + \boldsymbol{a}',$$

where k is a real number and $\boldsymbol{a}' \perp \mathbf{1}$. We introduce in the algebra an operation of conjugation whose effect on an element $\boldsymbol{a}$ is given by

$$\bar{\boldsymbol{a}} = k\mathbf{1} - \boldsymbol{a}'.$$

In particular, if $\boldsymbol{a}$ is proportional to $\mathbf{1}$, then $\bar{\boldsymbol{a}} = \boldsymbol{a}$, and if $\boldsymbol{a}$ is orthogonal to $\mathbf{1}$, then $\bar{\boldsymbol{a}} = -\boldsymbol{a}$. Clearly,

$$\bar{\bar{\boldsymbol{a}}} = \boldsymbol{a}$$

and

$$\overline{\boldsymbol{a} + \boldsymbol{b}} = \bar{\boldsymbol{a}} + \bar{\boldsymbol{b}}.$$

Now we are ready to present the ideas underlying the proof of our theorem.

Let $\mathcal{U}$ *be a subalgebra of the algebra* $\mathcal{A}$ *containing* $\mathbf{1}$ *and different from* $\mathcal{A}$. Let $\mathbf{1}, \boldsymbol{i}_1, \boldsymbol{i}_2, \ldots, \boldsymbol{i}_n$ be a basis of $\mathcal{U}$ such that $\boldsymbol{i}_1, \boldsymbol{i}_2, \ldots, \boldsymbol{i}_n$ are orthogonal to $\mathbf{1}$. Then the conjugate of an element $a_0\mathbf{1}+a_1\boldsymbol{i}_1+\ldots+a_n\boldsymbol{i}_n$ is the element $a_0\mathbf{1} - a_1\boldsymbol{i}_1 - \ldots - a_n\boldsymbol{i}_n$. This shows that if $\boldsymbol{u}$ is an element of $\mathcal{A}$, then so is its conjugate $\bar{\boldsymbol{u}}$.

According to chapter 12, there exists a nonzero vector orthogonal to $\mathcal{U}$. A suitable numerical multiple of it is a *unit vector* $\boldsymbol{e}$ *orthogonal to* $\mathcal{U}$. We shall show that the set of elements of the form

$$\boldsymbol{u}_1 + \boldsymbol{u}_2\boldsymbol{e} \quad (\boldsymbol{u}_1 \in \mathcal{U},\ \boldsymbol{u}_2 \in \mathcal{U}) \tag{17.2}$$

is closed under multiplication, and thus a subalgebra of $\mathcal{U}$. Let $\mathcal{U} + \mathcal{U}e$ denote this subalgebra. We shall prove that:

Assertion 17.1 *The representation of an element of* $\mathcal{U} + \mathcal{U}e$ *in the form* (17.2) *is unique;*

Assertion 17.2 *The product of two elements of the form* (17.2) *is given by*

$$(\boldsymbol{u}_1 + \boldsymbol{u}_2 \boldsymbol{e})(\boldsymbol{v}_1 + \boldsymbol{v}_2 \boldsymbol{e}) = (\boldsymbol{u}_1 \boldsymbol{v}_1 - \bar{\boldsymbol{v}}_2 \boldsymbol{u}_2) + (\boldsymbol{v}_2 \boldsymbol{u}_1 + \boldsymbol{u}_2 \bar{\boldsymbol{v}}_1)\boldsymbol{e}. \tag{17.3}$$

Juxtaposing these facts and the doubling procedure described in chapter 6 we arrive at the conclusion that *the subalgebra $\mathcal{U} + \mathcal{U}\boldsymbol{e}$ is isomorphic to the doubled subalgebra $\mathcal{U}$.*

The rest of the proof is relatively simple. Before turning to the last phase of the proof we remark on a certain aspect of conjugation in the algebra $\mathcal{A}$.

Since it contains an identity element $\mathbf{1}$, the algebra $\mathcal{A}$ contains the subalgebra of elements of the form $k\mathbf{1}$. This subalgebra is isomorphic to the algebra of real numbers. We denote it by $\mathcal{R}$. If in the preceding argument we replace $\mathcal{U}$ by $\mathcal{R}$, then $\boldsymbol{e}$ will be a unit vector orthogonal to $\mathbf{1}$. By formula (17.3)

$$\boldsymbol{e}^2 = (\mathbf{0} + 1\boldsymbol{e})(\mathbf{0} + 1\boldsymbol{e}) = -\mathbf{1}.$$

This implies that the square of a vector $\boldsymbol{a}'$ orthogonal to $\mathbf{1}$ is $\lambda\mathbf{1}$, where $\lambda \leq 0$. It is easy to show that, conversely, if the square of an element is $\lambda\mathbf{1}$ and $\lambda \leq 0$, then this element is orthogonal to $\mathbf{1}$.[5] Thus the elements orthogonal to $\mathbf{1}$, and only these elements, are characterized by the fact that their squares are equal to $\lambda\mathbf{1}$, where $\lambda \leq 0$. This enables us to give the following alternative description of conjugation in $\mathcal{A}$: *Let*

$$k\mathbf{1} + \boldsymbol{a}', \textit{ where } \boldsymbol{a}'^2 = \lambda\mathbf{1}, \quad \lambda \leq 0,$$

be the unique representation of an element $a \in A$. Then $\bar{\boldsymbol{a}} = k\mathbf{1} - \boldsymbol{a}'$.

We are now ready to present the last, quite transparent, part of the proof.

Consider once more the subalgebra $\mathcal{R}$. If $\mathcal{R} \neq \mathcal{A}$, then there is a unit vector $\boldsymbol{e}$ orthogonal to $\mathcal{R}$. Consider the subalgebra $\mathcal{C} = \mathcal{R} + \mathcal{R}\boldsymbol{e}$. Since it is the doubled algebra $\mathcal{R}$, it is isomorphic to the algebra of complex numbers. From what was said about conjugation in the algebra $\mathcal{A}$ it follows that for the elements of $\mathcal{C}$ conjugation coincides with the usual conjugation of complex numbers.

If the subalgebra $\mathcal{C}$ does not coincide with $\mathcal{A}$, then we can once more find a unit vector $\boldsymbol{e}'$ orthogonal to $\mathcal{C}$. We consider the subalgebra $\mathcal{Q} = \mathcal{C} + \mathcal{C}\boldsymbol{e}'$, the result of doubling $\mathcal{C}$. This algebra is isomorphic to the algebra of quaternions. Our earlier characterization of conjugation in $\mathcal{A}$ implies that for the elements of $\mathcal{Q}$ conjugation coincides with conjugation in the algebra of quaternions.

If the subalgebra $\mathcal{Q}$ is not all of $\mathcal{A}$, then we again choose a vector $\boldsymbol{e}''$ orthogonal to $\mathcal{Q}$ and consider the subalgebra $\mathcal{O} = \mathcal{Q} + \mathcal{Q}\boldsymbol{e}''$ which is the result of doubling $\mathcal{Q}$ and is therefore isomorphic to the Cayley numbers (chapter 6). This algebra *must* coincide with $\mathcal{A}$ for, as we shall show, *any subalgebra containing* $\mathbf{1}$ *and not equal to* $\mathcal{A}$ *is associative.* Since multiplication of Cayley numbers is not associative, the subalgebra $\mathcal{O}$ must coincide with the whole algebra $\mathcal{A}$.

In turn, if the algebra $\mathcal{A}$ is not isomorphic to one of the algebras $\mathcal{R}, \mathcal{C}, \mathcal{Q}$, then it is isomorphic to the algebra $\mathcal{O}$. But this is the assertion of our theorem.

We see that *our theorem will have been proved if we prove the assertions* 17.1 *and* 17.2, *and the assertion:*

Assertion 17.3 *Every subalgebra containing* $\mathbf{1}$ *and different from* $\mathcal{A}$ *is associative.*

17.3 Two Lemmas

First we establish two lemmas. We suggest that the reader familiarize himself with the statements and leave their proofs to a second reading.

Lemma 17.1 *the following identity holds in any normed algebra:*

$$(\boldsymbol{a}_1\boldsymbol{b}_1, \boldsymbol{a}_2\boldsymbol{b}_2) + (\boldsymbol{a}_1\boldsymbol{b}_2, \boldsymbol{a}_2\boldsymbol{b}_1) = 2(\boldsymbol{a}_1, \boldsymbol{a}_2)(\boldsymbol{b}_1, \boldsymbol{b}_2). \tag{17.4}$$

We note that this identity connects four elements $\boldsymbol{a}_1, \boldsymbol{a}_2, \boldsymbol{b}_1, \boldsymbol{b}_2$ of the algebra $\mathcal{A}$.

Proof. Put for $\boldsymbol{a}$ in the fundamental identity (17.1) the sum $\boldsymbol{a}_1 + \boldsymbol{a}_2$. We have

$$(\boldsymbol{a}_1\boldsymbol{b} + \boldsymbol{a}_2\boldsymbol{b}, \quad \boldsymbol{a}_1\boldsymbol{b} + \boldsymbol{a}_2\boldsymbol{b}) = (\boldsymbol{a}_1 + \boldsymbol{a}_2, \quad \boldsymbol{a}_1 + \boldsymbol{a}_2)(\boldsymbol{b}, \boldsymbol{b}),$$

or

$$\begin{aligned}(\boldsymbol{a}_1\boldsymbol{b}, \boldsymbol{a}_1\boldsymbol{b}) &+ (\boldsymbol{a}_2\boldsymbol{b}, \boldsymbol{a}_2\boldsymbol{b}) + 2(\boldsymbol{a}_1\boldsymbol{b}, \boldsymbol{a}_2\boldsymbol{b}) \\ &= (\boldsymbol{a}_1, \boldsymbol{a}_1)(\boldsymbol{b}, \boldsymbol{b}) + (\boldsymbol{a}_2, \boldsymbol{a}_2)(\boldsymbol{b}, \boldsymbol{b}) + 2(\boldsymbol{a}_1, \boldsymbol{a}_2)(\boldsymbol{b}, \boldsymbol{b}).\end{aligned}$$

By the fundamental identity, the first and second terms on the left are equal, respectively, to the first and second terms on the right. Hence

$$(\boldsymbol{a}_1\boldsymbol{b}, \boldsymbol{a}_2\boldsymbol{b}) = (\boldsymbol{a}_1, \boldsymbol{a}_2)(\boldsymbol{b}, \boldsymbol{b}). \tag{17.5}$$

To obtain the required result we must replace $\boldsymbol{b}$ in (17.5) by $\boldsymbol{b}_1 + \boldsymbol{b}_2$. Then we have

$$(\boldsymbol{a}_1\boldsymbol{b}_1 + \boldsymbol{a}_1\boldsymbol{b}_2, \boldsymbol{a}_2\boldsymbol{b}_1 + \boldsymbol{a}_2\boldsymbol{b}_2) = (\boldsymbol{a}_1, \boldsymbol{a}_2)(\boldsymbol{b}_1 + \boldsymbol{b}_2, \boldsymbol{b}_1 + \boldsymbol{b}_2),$$

or

$$(\boldsymbol{a}_1\boldsymbol{b}_1, \boldsymbol{a}_2\boldsymbol{b}_1) + (\boldsymbol{a}_1\boldsymbol{b}_2, \boldsymbol{a}_2\boldsymbol{b}_2) + (\boldsymbol{a}_1\boldsymbol{b}_1, \boldsymbol{a}_2\boldsymbol{b}_2) + (\boldsymbol{a}_1\boldsymbol{b}_2, \boldsymbol{a}_2\boldsymbol{b}_1)$$
$$= (\boldsymbol{a}_1, \boldsymbol{a}_2)(\boldsymbol{b}_1, \boldsymbol{b}_1) + (\boldsymbol{a}_1, \boldsymbol{a}_2)(\boldsymbol{b}_2, \boldsymbol{b}_2) + 2(\boldsymbol{a}_1, \boldsymbol{a}_2)(\boldsymbol{b}_1, \boldsymbol{b}_2).$$

By (17.5), the first and second summands on the left are equal, respectively, to the first and second summands on the right. Cancellation yields the identity (17.4).

Lemma 17.2 *The following identity holds in a normed algebra with identity:*

$$(\boldsymbol{a}\boldsymbol{b})\bar{\boldsymbol{b}} = (\boldsymbol{b}, \boldsymbol{b})\boldsymbol{a}. \tag{17.6}$$

In other words, the element $(\boldsymbol{a}\boldsymbol{b})\bar{\boldsymbol{b}}$ is always proportional to $\boldsymbol{a}$ and the proportionality coefficient is $(\boldsymbol{b}, \boldsymbol{b})$.

Proof. First we note that it suffices to prove the identity (17.6) for the case when $\boldsymbol{b} \perp \mathbf{1}$. Indeed, let $\boldsymbol{b}'$ be an element of the algebra $\mathcal{A}$. If we represent it in the form

$$\boldsymbol{b}' = k\mathbf{1} + \boldsymbol{b},$$

with $\boldsymbol{b} \perp \mathbf{1}$, then $\bar{\boldsymbol{b}} = -\boldsymbol{b}$, and

$$(\boldsymbol{a}\boldsymbol{b}')\bar{\boldsymbol{b}}' = (\boldsymbol{a}(k\mathbf{1} + \boldsymbol{b}))(k\mathbf{1} - \boldsymbol{b}) = k^2\boldsymbol{a} - (\boldsymbol{a}\boldsymbol{b})\boldsymbol{b} = k^2\boldsymbol{a} + (\boldsymbol{a}\boldsymbol{b})\bar{\boldsymbol{b}}.$$

If we assume that formula (17.6) holds for the vector $\boldsymbol{b}$, then we have

$$(\boldsymbol{a}\boldsymbol{b}')\bar{\boldsymbol{b}}' = k^2\boldsymbol{a} + (\boldsymbol{b}, \boldsymbol{b})\boldsymbol{a} = [k^2 + (\boldsymbol{b}, \boldsymbol{b})]\boldsymbol{a} = (\boldsymbol{b}', \boldsymbol{b}')\boldsymbol{a},^{6}$$

that is, formula (17.6) holds for $\boldsymbol{b}'$.

Thus we shall prove (17.6) under the assumption that $\boldsymbol{b} \perp \mathbf{1}$ (or, equivalently, $\bar{\boldsymbol{b}} = -\boldsymbol{b}$). Also, we shall write λ for $(\boldsymbol{b}, \boldsymbol{b})$.

Consider the element

$$\boldsymbol{c} = (\boldsymbol{a}\boldsymbol{b})\bar{\boldsymbol{b}} - \lambda\boldsymbol{a}.$$

We must show that $\boldsymbol{c} = \mathbf{o}$ or, equivalently, that

$$(\boldsymbol{c}, \boldsymbol{c}) = 0.$$

In view of the properties of scalar products we have

$$(\boldsymbol{c}, \boldsymbol{c}) = ((\boldsymbol{a}, \boldsymbol{b})\bar{\boldsymbol{b}}, (\boldsymbol{a}\boldsymbol{b})\bar{\boldsymbol{b}}) + \lambda^2(\boldsymbol{a}, \boldsymbol{a}) - 2\lambda((\boldsymbol{a}\boldsymbol{b})\bar{\boldsymbol{b}}, \boldsymbol{a}). \tag{17.7}$$

The right side is a sum of three terms. Using the fundamental identity (17.1) we can easily simplify the first summand:

$$((\boldsymbol{a}\boldsymbol{b})\bar{\boldsymbol{b}}, (\boldsymbol{a}\boldsymbol{b})\bar{\boldsymbol{b}}) = (\boldsymbol{a}\boldsymbol{b}, \boldsymbol{a}\boldsymbol{b})(\bar{\boldsymbol{b}}, \bar{\boldsymbol{b}}) = (\boldsymbol{a}, \boldsymbol{a})(\boldsymbol{b}, \boldsymbol{b})^2 = \lambda^2(\boldsymbol{a}, \boldsymbol{a}).$$

To simplify the third summand we use the identity (17.4). First we write it as

$$(\boldsymbol{a}_1\boldsymbol{b}_1, \boldsymbol{a}_2\boldsymbol{b}_2) = 2(\boldsymbol{a}_1, \boldsymbol{a}_2)(\boldsymbol{b}_1, \boldsymbol{b}_2) - (\boldsymbol{a}_1\boldsymbol{b}_2, \boldsymbol{a}_2\boldsymbol{b}_1).$$

In the last identity we put

$$\boldsymbol{a}_1 = \boldsymbol{ab}, \quad \boldsymbol{b}_1 = \bar{\boldsymbol{b}}, \quad \boldsymbol{a}_2 = \boldsymbol{a}, \quad \boldsymbol{b}_2 = \mathbf{1},$$

and obtain

$$((\boldsymbol{ab})\bar{\boldsymbol{b}}, \boldsymbol{a}) = 2(\boldsymbol{ab}, \boldsymbol{a})(\bar{\boldsymbol{b}}, \mathbf{1}) - (\boldsymbol{ab}, \boldsymbol{a}\bar{\boldsymbol{b}}).$$

Since $\boldsymbol{b} \perp \mathbf{1}$, the first summand on the right is zero, and the second is

$$-(\boldsymbol{ab}, \boldsymbol{a}\bar{\boldsymbol{b}}) = (\boldsymbol{ab}, \boldsymbol{ab}) = (\boldsymbol{a}, \boldsymbol{a})(\boldsymbol{b}, \boldsymbol{b}) = \lambda(\boldsymbol{a}, \boldsymbol{a}).$$

Hence

$$((\boldsymbol{ab})\bar{\boldsymbol{b}}, \boldsymbol{a}) = \lambda(\boldsymbol{a}, \boldsymbol{a}).$$

Now we can rewrite (17.7) and obtain

$$(\boldsymbol{c}, \boldsymbol{c}) = \lambda^2(\boldsymbol{a}, \boldsymbol{a}) + \lambda^2(\boldsymbol{a}, \boldsymbol{a}) - 2\lambda^2(\boldsymbol{a}, \boldsymbol{a}) = 0,$$

which is what we wished to prove.

A consequence of Lemma 17.2. We now deduce from the identity (17.6) another identity that will play a very important role in what follows.

If we replace $\boldsymbol{b}$ in (17.6) by $\boldsymbol{x} + \boldsymbol{y}$, then we obtain

$$(\boldsymbol{a}(\boldsymbol{x} + \boldsymbol{y}))(\bar{\boldsymbol{x}} + \bar{\boldsymbol{y}}) = (\boldsymbol{x} + \boldsymbol{y}, \boldsymbol{x} + \boldsymbol{y})\boldsymbol{a},$$

or

$$\begin{aligned}(\boldsymbol{ax})\bar{\boldsymbol{x}} + (\boldsymbol{ay})\bar{\boldsymbol{y}} + (\boldsymbol{ax})\bar{\boldsymbol{y}} + (\boldsymbol{ay})\bar{\boldsymbol{x}} \\ = (\boldsymbol{x}, \boldsymbol{x})\boldsymbol{a} + (\boldsymbol{y}, \boldsymbol{y})\boldsymbol{a} + 2(\boldsymbol{x}, \boldsymbol{y})\boldsymbol{a}.\end{aligned}$$

In view of (17.6), the first and second summands on the left are equal, respectively, to the first and second summands on the right. Hence

$$(\boldsymbol{ax})\bar{\boldsymbol{y}} + (\boldsymbol{ay})\bar{\boldsymbol{x}} = 2(\boldsymbol{xy})\boldsymbol{a}. \tag{17.8}$$

This is the identity we wished to establish.

▷ Putting $\boldsymbol{a} = \mathbf{1}$ in (17.6) we obtain

$$\boldsymbol{b}\bar{\boldsymbol{b}} = (\boldsymbol{b}, \boldsymbol{b})\mathbf{1}.$$

This and (17.6) yield

$$(\boldsymbol{ab})\bar{\boldsymbol{b}} = \boldsymbol{a}(\boldsymbol{b}\bar{\boldsymbol{b}}).$$

Hence

$$(\boldsymbol{ab})\boldsymbol{b} = \boldsymbol{a}(\boldsymbol{bb}).$$

A similar argument proves that

$$\boldsymbol{b}(\boldsymbol{ba}) = (\boldsymbol{bb})\boldsymbol{a}.$$

The last two formulas show that the algebra $\mathcal{A}$ is alternative. ◁

17.4 Conclusion of the Proof

It remains to prove the assertions 17.1, 17.2, and 17.3. We recall that there $\mathcal{U}$ denotes a subalgebra of the algebra $\mathcal{A}$ that contains $\mathbf{1}$ and does not coincide with $\mathcal{A}$, and $\boldsymbol{e}$ is a unit vector orthogonal to $\mathcal{U}$.

First we show that the subspaces $\mathcal{U}$ and $\mathcal{U}\boldsymbol{e}$ are orthogonal, that is, $\boldsymbol{u}_1 \perp \boldsymbol{u}_2\boldsymbol{e}$ for any two elements $\boldsymbol{u}_1 \in \mathcal{U}, \boldsymbol{u}_2 \in \mathcal{U}$.

We use lemma 17.1. If we put in (17.4) $\boldsymbol{a}_1 = \boldsymbol{u}_1$, $\boldsymbol{b}_1 = \boldsymbol{u}_2$, $\boldsymbol{a}_2 = \boldsymbol{e}$, $\boldsymbol{b}_2 = \mathbf{1}$, then we obtain

$$(\boldsymbol{u}_1\boldsymbol{u}_2, \boldsymbol{e}) + (\boldsymbol{u}_1, \boldsymbol{u}_2\boldsymbol{e}) = 2(\boldsymbol{u}_1, \boldsymbol{e})(\boldsymbol{u}_2, \mathbf{1}).$$

Now we need only bear in mind that $\mathcal{U}$ is a subalgebra, so that $\boldsymbol{u}_1\boldsymbol{u}_2$ is in $\mathcal{U}$. But then $\boldsymbol{u}_1 \perp \boldsymbol{e}, \boldsymbol{u}_1\boldsymbol{u}_2 \perp \boldsymbol{e}$. It now follows from the last equality that

$$(\boldsymbol{u}_1, \boldsymbol{u}_2\boldsymbol{e}) = 0,$$

that is, $\boldsymbol{u}_1 \perp \boldsymbol{u}_2\boldsymbol{e}$. This means that the subspaces $\mathcal{U}$ and $\mathcal{U}\boldsymbol{e}$ are orthogonal, as claimed.

Now we easily prove assertion 17.1: *The representation of any element in $\mathcal{U} + \mathcal{U}\boldsymbol{e}$ in the form $\boldsymbol{u}_1 + \boldsymbol{u}_2\boldsymbol{e}$ is unique.* In fact, suppose that

$$\boldsymbol{u}_1 + \boldsymbol{u}_2\boldsymbol{e} = \boldsymbol{u}'_1 + \boldsymbol{u}'_2\boldsymbol{e}.$$

Then

$$\boldsymbol{u}_1 - \boldsymbol{u}'_1 = (\boldsymbol{u}'_2 - \boldsymbol{u}_2)\boldsymbol{e}.$$

This means that the element $\boldsymbol{v} = \boldsymbol{u}_1 - \boldsymbol{u}'_1$ is in the subspaces $\mathcal{U}$ and $\mathcal{U}\boldsymbol{e}$. Since these subspaces have just been shown to be orthogonal, $(\boldsymbol{v}, \boldsymbol{v}) = 0$, and therefore $\boldsymbol{v} = \mathbf{o}$. This implies that $\boldsymbol{u}_1 - \boldsymbol{u}'_1 = \mathbf{o}$, and $(\boldsymbol{u}'_2 - \boldsymbol{u}_2)\boldsymbol{e} = \mathbf{o}$. Also, in view of the fundamental identity (17.1), $\boldsymbol{ab} = \mathbf{o}$ implies that $\boldsymbol{a} = \mathbf{o}$ or $\boldsymbol{b} = \mathbf{o}$. In our case $(\boldsymbol{u}'_2 - \boldsymbol{u}_2)\boldsymbol{e} = \mathbf{o}$ and $\boldsymbol{e} \neq \mathbf{o}$ imply that $\boldsymbol{u}'_2 - \boldsymbol{u}_2 = \mathbf{o}$. Hence $\boldsymbol{u}_1 = \boldsymbol{u}'_1$ and $\boldsymbol{u}_2 = \boldsymbol{u}'_2$. This completes the proof of assertion 17.1.

Next we prove assertion 17.2, that is, the correctness of formula (17.3). To this end we shall prove that if $\boldsymbol{u}$ and $\boldsymbol{v}$ are elements of the subalgebra $\mathcal{U}$, then

$$(\boldsymbol{ue})\boldsymbol{v} = (\boldsymbol{u}\bar{\boldsymbol{v}})\boldsymbol{e}, \tag{α}$$

$$u(ve) = (vu)e, \tag{β}$$

$$(ue)(ve) = -\bar{v}u. \tag{γ}$$

With these relations at our disposal we can easily prove formula (17.3). In fact,

$$(u_1 + u_2e)(v_1 + v_2e) = u_1v_1 + (u_2e)(v_2e) + (u_2e)v_1 + u_1(v_2e).$$

If we transform the last three terms on the right in accordance with the formulas $(\alpha), (\beta)$, and (γ), then we obtain the equality

$$(u_1 + u_2e)(v_1 + v_2e) = (u_1v_1 - \bar{v}_2u_2) + (v_2u_1 + u_2\bar{v}_1)e,$$

that is, formula (17.3).

To prove $(\alpha), (\beta)$, and (γ) we make use of the identity (17.8):

$$(ax)\bar{y} + (ay)\bar{x} = 2(x, y)a. \tag{17.8}$$

If we put in this identity

$$a = u, \quad x = e, \quad y = \bar{v}$$

and bear in mind that $\bar{v} \perp e$, then we have

$$(ue)v + (u\bar{v})\bar{e} = o.$$

Since $\bar{e} = -e$(for $e \perp 1$), we obtain the formula (α).

To prove (β), put in (17.8)

$$a = 1, \quad x = u, \quad y = \overline{ve}.$$

Since $\overline{ve} = -ve \quad (ve \perp \mathcal{U}$, so that $ve \perp 1$), it follows that

$$u(ve) - (ve)\bar{u} = o.$$

Using (α) we obtain

$$u(ve) = (ve)\bar{u} = (vu)e.$$

To prove (γ) we use the following obvious remark: If this formula holds for $v = c$ and $v = d$, then it also holds for $v = c + d$. Since every element v can be written as a sum of two terms one of which is proportional to 1 and the other orthogonal to 1, it suffices to prove (γ) in two cases: when $v = k1$ and when $v \perp 1$.

If $\boldsymbol{v} = k\mathbf{1}$, then formula (γ) becomes

$$k(\boldsymbol{ue})\boldsymbol{e} = -k\boldsymbol{u},$$

an identity whose validity is implied by the identity (17.6).

Now suppose that $\boldsymbol{v} \perp \mathbf{1}$ (so that $\bar{\boldsymbol{v}} = -\boldsymbol{v}$). If in (17.8) we put

$$\boldsymbol{a} = \boldsymbol{u}, \quad \boldsymbol{x} = \boldsymbol{e}, \quad \boldsymbol{y} = -\boldsymbol{ve},$$

then we have

$$(\boldsymbol{ue})(\boldsymbol{ve}) - (\boldsymbol{u}(\boldsymbol{ve}))\bar{\boldsymbol{e}} = -2(\boldsymbol{e}, \boldsymbol{ve})\boldsymbol{u}.$$

By the identity (17.5), $(\boldsymbol{e}, \boldsymbol{ve})$ equals $(\mathbf{1}, \boldsymbol{v})(\boldsymbol{e}, \boldsymbol{e})$, that is, zero. Further, by (β), the second term on the left equals $-((\boldsymbol{vu})\boldsymbol{e})\bar{\boldsymbol{e}} = -\boldsymbol{vu} = \bar{\boldsymbol{v}}\boldsymbol{u}$. But then

$$(\boldsymbol{ue})(\boldsymbol{ve}) = -\bar{\boldsymbol{v}}\boldsymbol{u},$$

which is what we wished to prove. By proving $(\alpha), (\beta), (\gamma)$ we have proved assertion 17.2.

In order to complete the proof of our theorem we must prove assertion 17.3: *Every subalgebra $\mathcal{U}$ of the algebra $\mathcal{A}$ that contains $\mathbf{1}$ and is not $\mathcal{A}$ is associative*, that is,

$$(\boldsymbol{uv})\boldsymbol{w} = \boldsymbol{u}(\boldsymbol{vw})$$

for any three elements $\boldsymbol{u}, \boldsymbol{v}, \boldsymbol{w}$ in $\mathcal{U}$.

To show this we again use (17.8). Putting in (17.8)

$$\boldsymbol{a} = \boldsymbol{ve}, \quad \boldsymbol{x} = \bar{\boldsymbol{w}}, \quad \boldsymbol{y} = \bar{\boldsymbol{u}}\boldsymbol{e},$$

we have

$$((\boldsymbol{ve})\bar{\boldsymbol{w}})(-\bar{\boldsymbol{u}}\boldsymbol{e}) + ((\boldsymbol{ve})(\bar{\boldsymbol{u}}\boldsymbol{e}))\boldsymbol{w} = \mathbf{o},$$

or, using (α) and (γ),

$$\boldsymbol{u}(\boldsymbol{vw}) - (\boldsymbol{uv})\boldsymbol{w} = \mathbf{o}.$$

This completes the proof of Hurwitz's theorem.

Chapter 18

A Method for Constructing All Normed Algebras and Its Implications for the Problem of the Sum of Squares

18.1 A Method for Constructing New Normed Algebras

First we describe a special method for constructing many normed algebras starting with a normed algebra $\mathcal{A}$.

Let A and B be two orthogonal (that is, norm-preserving) transformations on $\mathcal{A}$. In the vector space $\mathcal{A}$ we define a new multiplication $\circ$ given by the formula

$$\boldsymbol{u} \circ \boldsymbol{v} = A(\boldsymbol{u})B(\boldsymbol{v}). \tag{18.1}$$

This definition states that the new product of two elements $\boldsymbol{u}$ and $\boldsymbol{v}$ is equal to the old product of their transforms $A(\boldsymbol{u})$ and $B(\boldsymbol{v})$.

It is easy to see that the new operation satisfies the following relations:

$$\boldsymbol{u} \circ (\boldsymbol{v}_1 + \boldsymbol{v}_2) = \boldsymbol{u} \circ \boldsymbol{v}_1 + \boldsymbol{u} \circ \boldsymbol{v}_2, \quad \boldsymbol{u} \circ k\boldsymbol{v} = k(\boldsymbol{u} \circ \boldsymbol{v}),$$

and

$$(\boldsymbol{u}_1 + \boldsymbol{u}_2) \circ \boldsymbol{v} = \boldsymbol{u}_1 \circ \boldsymbol{v} + \boldsymbol{u}_2 \circ \boldsymbol{v}, \quad k\boldsymbol{u} \circ \boldsymbol{v} = k(\boldsymbol{u} \circ \boldsymbol{v}).$$

The first two of these relations are implied by the linearity of the transformation B and the last two by the linearity of the transformation A. These relations show that the new operation is indeed a multiplication (see section 7.7).

The vector space of the algebra $\mathcal{A}$ with the new multiplication is denoted by $\mathcal{A}_o$. Thus $\mathcal{A}$ and $\mathcal{A}_o$ are copies of the same vector space furnished with different multiplications.

The algebra $\mathcal{A}$ is furnished with a scalar product $(\boldsymbol{x}, \boldsymbol{y})$. It turns out that, like the old algebra $\mathcal{A}$, the new algebra $\mathcal{A}_o$ is normed with respect to this scalar product. In fact, formula (18.1) implies that

$$|\boldsymbol{u} \circ \boldsymbol{v}| = |A(\boldsymbol{u})B(\boldsymbol{v})| = |A(\boldsymbol{u})||B(\boldsymbol{v})| = |\boldsymbol{u}||\boldsymbol{v}|;$$

here we made use of the fact that the original algebra $\mathcal{A}$ is normed and the transformations A and B are orthogonal, that is, that

$$|A(\boldsymbol{u})| = |\boldsymbol{u}| \text{ and } |B(\boldsymbol{v})| = |\boldsymbol{v}|.$$

18.2 Construction of All Normed Algebras

The above method enables us to obtain many normed algebras from a given normed algebra. For this we need only substitute in formula (18.1) different pairs of orthogonal transformations. We are familiar with four remarkable normed algebras: the real numbers, the complex numbers, the quaternions, and the Cayley numbers. It is not unreasonable to ask whether all normed algebras can be obtained from these four by means of the method just described. It turns out that the answer to this question is affirmative. Since, by Hurwitz's theorem, these four algebras are the only normed algebras with an identity, we must prove the following theorem.

Theorem 18.1 *Every normed algebra $\mathcal{A}_o$ can be obtained from a normed algebra $\mathcal{A}$ with an identity by the introduction of a new multiplication via formula* (18.1) (*in that formula* $\circ$ *denotes the multiplication in the algebra* $\mathcal{A}_o$.)

For proof take an element $\boldsymbol{e}$ of norm 1 in $\mathcal{A}_o$, and consider the transformation that maps an element $\boldsymbol{x}$ in $\mathcal{A}_o$ onto $\boldsymbol{x} \circ \boldsymbol{e}$. Such a transformation (multiplication by $\boldsymbol{e}$ on the right) is called a *right translation by* $\boldsymbol{e}$ in the algebra $\mathcal{A}_o$. We denote it by A. Thus

$$A(\boldsymbol{x}) = \boldsymbol{x} \circ \boldsymbol{e}.$$

It is easy to see that the transformation A is orthogonal. In fact, the equalities

$$|A(\boldsymbol{x})| = |\boldsymbol{x} \circ \boldsymbol{e}| = |\boldsymbol{x}||\boldsymbol{e}| = |\boldsymbol{x}|$$

show that the transformation A preserves the norm of every element $\boldsymbol{x}$.

Similarly, we can introduce a second transformation

$$B(\boldsymbol{x}) = \boldsymbol{e} \circ \boldsymbol{x}$$

– *a left translation by an element* $\boldsymbol{e}$ in the algebra $\mathcal{A}_o$ – and show that it is also orthogonal.

The orthogonality of the transformations A and B implies (see chapter 13) the existence of inverse transformations A^{-1} and B^{-1} as well as their orthogonality.[7]

We use these transformations to introduce a new multiplication in the vector space of the algebra $\mathcal{A}_o$. If $\boldsymbol{x}$ and $\boldsymbol{y}$ are two elements in $\mathcal{A}_o$, then their new product is defined by the formula

$$\boldsymbol{x}\boldsymbol{y} = A^{-1}(\boldsymbol{x}) \circ B^{-1}(\boldsymbol{y}). \qquad (18.2)$$

We denote the resulting algebra by $\mathcal{A}$.

The equality (18.2) expresses the new multiplication in term of the old. But we can easily use (18.2) to express the old multiplication in terms of the new: Putting

$$A^{-1}(\boldsymbol{x}) = \boldsymbol{u}, \quad B^{-1}(\boldsymbol{y}) = \boldsymbol{v},$$

we have

$$A(\boldsymbol{u})B(\boldsymbol{v}) = \boldsymbol{u} \circ \boldsymbol{v}.$$

Thus if we regard $\mathcal{A}$ with the operation $\boldsymbol{u}\boldsymbol{v}$ as the initial algebra, then the algebra $\mathcal{A}_o$, given from the very beginning, can be obtained from it by replacing its multiplication by the new multiplication $\boldsymbol{u} \circ \boldsymbol{v}$ in accordance with the formula

$$\boldsymbol{u} \circ \boldsymbol{v} = A(\boldsymbol{u})B(\boldsymbol{v}).$$

To complete the proof we need only show that the algebra $\mathcal{A}$ has an identity.

We claim that the element $\tilde{e} = \boldsymbol{e} \circ \boldsymbol{e}$ plays the role of the identity in the algebra $\mathcal{A}$. In fact, consider the products

$$\boldsymbol{x}\tilde{e} \text{ and } \tilde{e}\boldsymbol{y}.$$

For the first of these products we have

$$\boldsymbol{x}\tilde{e} = A^{-1}(\boldsymbol{x}) \circ B^{-1}(\tilde{e}).$$

By definition of the inverse transformation, the element $\boldsymbol{u} = B^{-1}(\tilde{e})$ is the (unique) solution of the equation

$$B(\boldsymbol{u}) = \tilde{e},$$

or, equivalently, of the equation $\boldsymbol{e} \circ \boldsymbol{u} = \boldsymbol{e} \circ \boldsymbol{e}$. From this it follows (by uniqueness) that this element is $\boldsymbol{e}$. Further, $\boldsymbol{v} = A^{-1}(\boldsymbol{x})$ means that $\boldsymbol{x} = A(\boldsymbol{v})$, that is, $\boldsymbol{v} \circ \boldsymbol{e} = \boldsymbol{x}$. But then

$$\boldsymbol{x}\tilde{e} = A^{-1}(\boldsymbol{x}) \circ B^{-1}(\tilde{e}) = \boldsymbol{v} \circ \boldsymbol{e} = \boldsymbol{x}.$$

Similarly,

$$\tilde{e}\boldsymbol{y} = A^{-1}(\tilde{e}) \circ B^{-1}(\boldsymbol{y}) = \boldsymbol{e} \circ B^{-1}(\boldsymbol{y}) = \boldsymbol{y}.$$

We have shown that $\tilde{e}$ is the identity of the algebra $\mathcal{A}$. This completes the proof of our theorem.

To recapitulate: *All normed algebras can be obtained from the four familiar algebras $\mathcal{R}, \mathcal{C}, \mathcal{Q}$, and $\mathcal{O}$ by the introduction of a new multiplication via formula* (18.1). In a sense, this may be viewed as a method for obtaining all normed algebras.

18.3 The Number n in the Identity (!)

One of the consequences of this theorem is that the dimension of any normed algebra is equal to one of the numbers 1,2,4,8 (these are the dimensions of the algebras of real numbers, complex numbers, quaternions, and Cayley numbers).

We note that there is a definite connection between normed algebras and identities of the form

$$(x_1^2 + x_2^2 + \ldots + x_n^2)(y_1^2 + y_2^2 + \ldots + y_n^2) \\ = \Phi_1^2 + \Phi_2^2 + \ldots + \Phi_n^2. \quad (!)$$

This connection (see chapter 17) consists in the fact that by taking a normed algebra, selecting in it an orthogonal basis, and writing down the multiplication rule in this basis, we obtain n forms $\Phi_1, \Phi, \ldots, \Phi_n$ satisfying the identity (!). Furthermore, all identities (!) can be obtained in this manner. In view of this connection we arrive at the following fundamental conclusion:

The number n in the identity (!)*can take on only the four values* $1, 2, 4$, *and* 8.

18.4 Survey of All Identities (!)

▷ From the theorem just proved we can deduce far more than just the number of squares in the identity (!). In fact, if we know how to construct all normed algebras, then we also have a method of describing all identities (!). We shall show that all such identities can be obtained in the following manner.

Choose one of the values 2, 4, 8 of n. In the corresponding one of the three algebras $\mathcal{C}, \mathcal{Q}, \mathcal{O}$ select three orthonormal bases

$$\boldsymbol{e}_1, \boldsymbol{e}_2, \ldots, \boldsymbol{e}_n; \quad \boldsymbol{k}_1, \boldsymbol{k}_2, \ldots, \boldsymbol{k}_n; \quad \boldsymbol{i}_1, \boldsymbol{i}_2, \ldots, \boldsymbol{i}_n.$$

If $\boldsymbol{x}$ and $\boldsymbol{y}$ are general vectors expressed in terms of the first and second of these bases, respectively, and their product $\boldsymbol{xy}$ is expressed in terms of the third basis, that is, if

$$\boldsymbol{x} = \sum_a x_a \boldsymbol{e}_a, \quad \boldsymbol{y} = \sum_\beta y_\beta \boldsymbol{k}_\beta, \quad \boldsymbol{xy} = \sum_\gamma \Phi_\gamma \boldsymbol{i}_\gamma,$$

then the forms $\Phi_\gamma(x_1, \ldots, x_n; \ y_1, \ldots, y_n)$ satisfy the identity (!).[8] *Moreover, all identities* (!) *can be obtained in this way.*

That our procedure always leads to forms satisfying (!) is apparent from the equality

$$(\sum_a x_a \boldsymbol{e}_a)(\sum_\beta y_\beta \boldsymbol{k}_\beta) = \sum_\gamma \Phi_\gamma \boldsymbol{i}_\gamma.$$

Taking norms on both sides we obtain an identity (!), as claimed.

Conversely, we shall show that all identities (!) can be obtained from multiplications of the form

$$\boldsymbol{x} \circ \boldsymbol{y} = A(\boldsymbol{x})B(\boldsymbol{y})$$

(A and B are orthogonal transformations) by expressing the products with respect to orthonormal bases $\boldsymbol{i}_1, \boldsymbol{i}_2, \ldots, \boldsymbol{i}_n$. In other words, the forms $\Phi_1, \Phi_2, \ldots, \Phi_n$ associated with a given identity come from an equality

$$A(\sum_\alpha x_\alpha \boldsymbol{i}_\alpha)B(\sum_\beta y_\beta \boldsymbol{i}_\beta) = \sum_\gamma \Phi_\gamma \boldsymbol{i}_\gamma. \tag{18.3}$$

The linearity of the transformations A and B implies that

$$A(\sum_\alpha x_\alpha \boldsymbol{i}_\alpha) = \sum_\alpha x_\alpha A(\boldsymbol{i}_\alpha), \quad B(\sum_\beta y_\beta \boldsymbol{i}_\beta) = \sum_\beta y_\beta B(\boldsymbol{i}_\beta).$$

Denote $A(\boldsymbol{i}_\alpha)$ by $\boldsymbol{e}_\alpha$ and $B(\boldsymbol{i}_\beta)$ by $\boldsymbol{k}_\beta$. Then (18.3) takes the form

$$(\sum_\alpha x_\alpha \boldsymbol{e}_\alpha)(\sum_\beta y_\beta \boldsymbol{k}_\beta) = \sum_\gamma \Phi_\gamma \boldsymbol{i}_\gamma.$$

Bearing in mind that the bases $\boldsymbol{e}_1, \boldsymbol{e}_2, \ldots, \boldsymbol{e}_n$ and $\boldsymbol{k}_1, \boldsymbol{k}_2, \ldots, \boldsymbol{k}_n$ are orthonormal, we conclude that the forms $\Phi_1, \Phi_2, \ldots, \Phi_n$ are obtained in the manner asserted in our theorem.

For mathematically more advanced readers the same survey of the identities (!) can be described differently. Given one of the permissible values 1,2,4,8 of n we choose n forms $\Phi_1, \Phi_2, \ldots, \Phi_n$ and alter them as follows: Replace the variables $x_1, x_2, \ldots, x_n$ in Φ_i by new variables $x'_1, x'_2, \ldots, x'_n$ related to them by an orthogonal transformation A. Apply a similar operation, involving another orthogonal transformation B, to the variables $y_1, y_2, \ldots, y_n$. Then apply to the forms $\Phi_1, \Phi_2, \ldots, \Phi_n$ a third orthogonal transformation C.

Call two identities (!) *equivalent* if they are obtained from one another in the indicated manner. Then we can say that, *for a given value of* $n = 1, 2, 4, 8$, *there is, up to equivalence, just one identity* (!). $\triangleleft$

18.5 Examples of 2- and 4-Dimensional Algebras and of the Associated Identities (!)

We know that there is just one normed 2-dimensional algebra with identity, namely, the algebra $\mathcal{C}$ of complex numbers. Since conjugation of complex numbers, that is, the mapping

$$\boldsymbol{x} \to \bar{\boldsymbol{x}},$$

is an orthogonal transformation of the algebra $\mathcal{C}$ (for $|\bar{\boldsymbol{x}}| = |\boldsymbol{x}|$), we can obtain at least three new algebras by replacing the usual multiplication of complex numbers with the multiplications

$$\begin{aligned} \boldsymbol{x} \textcircled{1} \boldsymbol{y} &= \bar{\boldsymbol{x}}\boldsymbol{y}, \\ \boldsymbol{x} \textcircled{2} \boldsymbol{y} &= \boldsymbol{x}\bar{\boldsymbol{y}}, \\ \boldsymbol{x} \textcircled{3} \boldsymbol{y} &= \bar{\boldsymbol{x}}\bar{\boldsymbol{y}}. \end{aligned} \tag{18.4}$$

In this way we obtain three new normed algebras $\mathcal{C}_1, \mathcal{C}_2, \mathcal{C}_3$.

It is a useful exercise for the reader to show that any normed 2-dimensional algebra is isomorphic to one of the algebras $\mathcal{C}, \mathcal{C}_1, \mathcal{C}_2, \mathcal{C}_3$ and no two of the latter are isomorphic.

We shall now give examples of identities (!) associated with these algebras.

Using the basis $\mathbf{1}, \boldsymbol{i}$ in the algebra $\mathcal{C}_1$ we have

$$\begin{aligned} \boldsymbol{x} \circled{1} \boldsymbol{y} &= \bar{\boldsymbol{x}}\boldsymbol{y} = (x_1 - x_2\boldsymbol{i})(y_1 + y_2\boldsymbol{i}) \\ &= (x_1y_1 + x_2y_2) + (x_1y_2 - x_2y_1)\boldsymbol{i}, \end{aligned}$$

so that the corresponding identity is

$$(x_1^2 + x_2^2)(y_1^2 + y_2^2) = (x_1y_1 + x_2y_2)^2 + (x_1y_2 - x_1y_1)^2.$$

We see that this identity is somewhat different from the familiar identity

$$(x_1^2 + x_2^2)(y_1^2 + y_2^2) = (x_1y_1 - x_2y_2)^2 + (x_1y_2 + x_2y_1)^2 \tag{18.5}$$

associated with the same basis $\mathbf{1}, \boldsymbol{i}$ in the algebra $\mathcal{C}$.

To obtain an identity that departs more radically from (18.5) we use the orthonormal basis (easily shown to be one)

$$\boldsymbol{e}_1 = \frac{1}{\sqrt{2}}(\mathbf{1} + \boldsymbol{i}) \text{ and } \boldsymbol{e}_2 = \frac{1}{\sqrt{2}}(\mathbf{1} - \boldsymbol{i})$$

in $\mathcal{C}$. Expressed in terms of this basis the multiplication rule takes the form

$$\begin{aligned} (u_1\boldsymbol{e}_1 + u_2\boldsymbol{e}_2)(v_1\boldsymbol{e}_1 + v_2\boldsymbol{e}_2) &= \frac{1}{\sqrt{2}}(u_1v_2 + v_1u_2 + u_1v_1 - u_2v_2)\boldsymbol{e}_1 \\ &+ \frac{1}{\sqrt{2}}(u_1v_2 + v_1u_2 - u_1v_1 + u_2v_2)\boldsymbol{e}_2. \end{aligned}$$

The corresponding identity is

$$\begin{aligned} (u_1^2 + u_2^2)(v_1^2 + v_2^2) &= [\frac{1}{\sqrt{2}}(u_1v_2 + v_1u_2 + u_1v_1 - u_2v_2)]^2 \\ &+ [\frac{1}{\sqrt{2}}(u_1v_2 + v_1u_2 - u_1v_1 + u_2v_2)]^2. \end{aligned}$$

Next we turn to 4-dimensional normed algebras. We know that, in this case, the only normed algebra with identity is the algebra $\mathcal{Q}$ of quaternions. Just as in the case of the complex numbers, conjugation effects an orthogonal transformation of the algebra $\mathcal{Q}$. Therefore, in addition to $\mathcal{Q}$, there are at least three more normed algebras $\mathcal{Q}_1$, $\mathcal{Q}_2$, $\mathcal{Q}_3$ with multiplications given by the formulas (18.4). In the 4-dimensional case, however, there are other normed algebras, such as the algebras with multiplication operations given by the formulas

$$\boldsymbol{axyb},$$
$$\boldsymbol{a\bar{x}yb},$$
$$\boldsymbol{ax\bar{y}b},$$
$$\boldsymbol{a\bar{x}\bar{y}b},$$

where $\boldsymbol{a}$ and $\boldsymbol{b}$ are two fixed quaternions.

We leave it to the reader to prove that every normed 4-dimensional algebra is isomorphic to an algebra of this type.

As an example, we consider the first of the above multiplications with $\boldsymbol{a} = \boldsymbol{i}$, $\boldsymbol{b} = \boldsymbol{j}$. We obtain an algebra $\tilde{Q}$ with the multiplication rule

$$\boldsymbol{x} \circ \boldsymbol{y} = (\boldsymbol{ix})(\boldsymbol{yj}).$$

We shall obtain the identity associated with this algebra in the basis $\mathbf{1}, \boldsymbol{i}, \boldsymbol{j}, \boldsymbol{k}$. We have

$$\begin{aligned}
\boldsymbol{x} \circ \boldsymbol{y} &= (\boldsymbol{i}(x_0 + x_1\boldsymbol{i} + x_2\boldsymbol{j} + x_3\boldsymbol{k}))((y_0 + y_1\boldsymbol{i} + y_2\boldsymbol{j} + y_3\boldsymbol{k})\boldsymbol{j}) \\
&= (x_0\boldsymbol{i} - x_1 + x_2\boldsymbol{k} - x_3\boldsymbol{j})(y_0\boldsymbol{j} + y_1\boldsymbol{k} - y_2 - y_3\boldsymbol{i}) \\
&= (x_1y_2 - x_2y_1 + x_0y_3 + x_3y_0) \\
&+ (-x_0y_2 - x_2y_0 + x_1y_3 - x_3y_1)\boldsymbol{i} \\
&+ (-x_1y_0 - x_0y_1 + x_3y_2 - x_2y_1)\boldsymbol{j} \\
&+ (x_0y_0 - x_1y_1 - x_2y_2 - x_3y_3)\boldsymbol{k}.
\end{aligned}$$

Hence the corresponding identity is

$$\begin{aligned}
&(x_0^2 + x_1^2 + x_2^2 + x_3^2)(y_0^2 + y_1^2 + y_2^2 + y_3^2) \\
&\quad = (x_1y_2 - x_2y_1 + x_0y_3 + x_3y_0)^2 + (-x_0y_2 - x_2y_0 + x_1y_3 - x_3y_1)^2 \\
&\quad +(-x_1y_0 - x_0y_1 + x_3y_2 - x_2y_1)^2 + (x_0y_0 - x_1y_1 - x_2y_2 - x_3y_3)^2.
\end{aligned}$$

We do not give examples of the identity (!) for $n = 8$ (other than the standard identity in section 6.6), for they are cumbersome and, if needed, can be obtained without essential difficulties.

Chapter 19

Frobenius' Theorem

19.1 Formulation of Frobenius' Theorem

One of the classical problems of the theory of algebras is that of finding all division algebras. In spite of the fundamental nature of the problem (and the fact that many problems in other areas of mathematics—such as topology—hinge on its solution), it is still not completely solved. An important result was obtained rather recently. It is to the effect that the dimension of such an algebra must be equal to one of the numbers 1, 2, 4, 8. While this shows that the dimensions of division algebras are small, we still have no complete overview of these algebras.

A considerably simpler problem is that of finding the division algebras satisfying additional natural conditions. Thus in 1878 the German mathematician Frobenius established the following remarkable result.

Frobenius' theorem. *Every associative division algebra is isomorphic to one of the following: the algebra of real numbers, the algebra of complex numbers, and the algebra of quaternions.*

Subsequently, the following, more general result, which may be called the generalized Frobenius theorem, was established. Its statement follows.

The generalized Frobenius theorem. *Every alternative division algebra is isomorphic to one of the following four algebras: the real numbers, the complex numbers, the quaternions, and the Cayley numbers.*

We recall that an algebra is alternative if the following identities hold for any two of its elements a and b:

$$(ab)b = a(bb),$$
$$(bb)a = b(ba).$$

It is clear that every associative algebra is alternative, so that Frobenius' theorem follows from the generalized Frobenius theorem. On the other hand, the algebra of Cayley numbers is alternative but not associative, so that the two theorems are actually different.

To prove these two theorems we first list certain properties of associative division algebras. Then we use these properties to prove Frobenius' theorem. This done, we give the proofs of these properties. In the last paragraph of this chapter we give a proof of the generalized Frobenius theorem based on Hurwitz's theorem.

19.2 Three Properties of Associative Division Algebras

Let $\mathcal{A}$ denote an associative division algebra. We claim that the algebra $\mathcal{A}$ has the following properties.

Assertion 19.1 *The algebra $\mathcal{A}$ has an identity.*

Assertion 19.2 *If an element $\boldsymbol{a} \in \mathcal{A}$ is not proportional to $\mathbf{1}$ then the set of elements C_a of the form*

$$\alpha\mathbf{1} + \beta\boldsymbol{a}$$

forms a subalgebra isomorphic to the algebra of complex numbers.

Assertion 19.3 *If two elements $\boldsymbol{a}_1 \in \mathcal{A}, \boldsymbol{a}_2 \in \mathcal{A}$ do not belong to the same subalgebra C_a then the set $\mathcal{Q}_{a_1,a_2}$ of elements of the form*

$$\alpha\mathbf{1} + \beta\boldsymbol{a}_1 + \gamma\boldsymbol{a}_2 + \delta\boldsymbol{a}_1\boldsymbol{a}_2$$

forms a subalgebra isomorphic to the algebra of quaternions.

In the process of proving assertion 19.3 we shall show that if $\boldsymbol{b}_1$ and $\boldsymbol{b}_2$ are two elements whose squares are $-\mathbf{1}$, then

$$\boldsymbol{b}_1\boldsymbol{b}_2 + \boldsymbol{b}_2\boldsymbol{b}_1 = \lambda\mathbf{1}, \tag{19.1}$$

where λ is a real number.

19.3 Proof of Frobenius' Theorem

Using properties of assertions 19.1, 19.2, and 19.3 it is an easy matter to prove Frobenius' theorem. Thus let $\mathcal{A}$ be an associative division algebra. By assertion 19.1, the algebra $\mathcal{A}$ has an identity. The elements of the form

$k\mathbf{1}$ form a subalgebra $\mathcal{R}$ isomorphic to the algebra of real numbers. If $\mathcal{R}$ is not all of $\mathcal{A}$ then, by assertion 19.2, $\mathcal{A}$ contains a subalgebra $\mathcal{C}_a$ isomorphic to the complex numbers. If $\mathcal{C}_a$ is not all of $\mathcal{A}$ then, by assertion 19.3, $\mathcal{A}$ contains a subalgebra $\mathcal{Q}_{a,b}$ isomorphic to the quaternion algebra. If $\mathcal{Q}_{a,b}$ coincides with $\mathcal{A}$, then we are done. Assume that this is not the case. Then $\mathcal{A}$ contains an element $\boldsymbol{c}$ not in $\mathcal{Q}_{a,b}$, and we shall show that $\mathcal{A}$ cannot be a division algebra.

In the quaternion algebra $\mathcal{Q}_{a,b}$ we choose a basis $\mathbf{1}, \boldsymbol{i}, \boldsymbol{j}, \boldsymbol{k}$ with the "standard" multiplication table

$$\boldsymbol{i}^2 = \boldsymbol{j}^2 = \boldsymbol{k}^2 = -\mathbf{1},$$
$$\boldsymbol{ij} = -\boldsymbol{ji} = \boldsymbol{k}, \quad \boldsymbol{jk} = -\boldsymbol{kj} = \boldsymbol{i}, \quad \boldsymbol{ki} = -\boldsymbol{ik} = \boldsymbol{j},$$

and write $\boldsymbol{c}$ as $p\mathbf{1} + q\boldsymbol{e}$, where $\boldsymbol{e}^2 = -\mathbf{1}$ ($\boldsymbol{e}$ is the "imaginary unit" of the complex algebra $\mathcal{C}_c$).

Next we rewrite the element $\boldsymbol{ie}$ using the associativity of $\mathcal{A}$ and relation (19.1). We have

$$\begin{aligned}\boldsymbol{ie} = (\boldsymbol{jk})\boldsymbol{e} = \boldsymbol{j}(\boldsymbol{ke}) = \boldsymbol{j}(-\boldsymbol{ek}+\lambda'\mathbf{1}) &= -(\boldsymbol{je})\boldsymbol{k}+\lambda'\boldsymbol{j}\\ &= -(-\boldsymbol{ej}+\lambda''\mathbf{1})\boldsymbol{k}+\lambda'\boldsymbol{j} = \boldsymbol{ei} - \lambda''\boldsymbol{k} + \lambda'\boldsymbol{j},\end{aligned}$$

and hence

$$\boldsymbol{ie} - \boldsymbol{ei} = \lambda'\boldsymbol{j} - \lambda''\boldsymbol{k}.$$

On the other hand, again by (19.1),

$$\boldsymbol{ie} + \boldsymbol{ei} = \lambda'''\mathbf{1}.$$

Adding the last two equalities we see that $\boldsymbol{ie}$ is an element of $\mathcal{Q}_{a,b}$. So is $\boldsymbol{ic} = \boldsymbol{i}(p\mathbf{1} + q\boldsymbol{e})$. If $\boldsymbol{c}' \in \mathcal{Q}_{a,b}$, then the product $\boldsymbol{ic}'$ is also an element of $\mathcal{Q}_{a,b}$. Thus the product of $\boldsymbol{i}$ by any element in $\mathcal{A}$ is an element of $\mathcal{Q}_{a,b}$. But this is impossible, for $\mathcal{A}$ is a division algebra (the equation $\boldsymbol{ix} = \boldsymbol{c}$, $\boldsymbol{c}$ not in $\mathcal{Q}_{a,b}$, is not solvable). This contradiction proves Frobenius' theorem.

It remains to prove assertions 19.1, 19.2, and 19.3.

19.4 Proof of the Three Assertions

Proof of assertion 19.1. Let $\boldsymbol{a}$ be a nonzero element of the algebra $\mathcal{A}$. We consider the equation

$$\boldsymbol{xa} = \boldsymbol{a}.$$

Since $\mathcal{A}$ is a division algebra, our equation has a unique solution $\boldsymbol{e}$, so that $\boldsymbol{ea} = \boldsymbol{a}$. Multiplying this equation on the left by $\boldsymbol{b}$ we obtain $\boldsymbol{b}(\boldsymbol{ea}) = \boldsymbol{ba}$, or, in view of the associativity of $\mathcal{A}$, $(\boldsymbol{be})\boldsymbol{a} = \boldsymbol{ba}$. Since the equation $\boldsymbol{xa} = \boldsymbol{ba}$ is uniquely solvable, it follows that

$$\boldsymbol{be} = \boldsymbol{b}.$$

If we multiply this relation on the right by $\boldsymbol{c}$ and argue in an analogous manner, then we find that

$$\boldsymbol{ec} = \boldsymbol{c}.$$

Since $\boldsymbol{b}$ and $\boldsymbol{c}$ are arbitrary elements, the last two equalities show that the element $\boldsymbol{e}$ is an identity of $\mathcal{A}$. As usual, we denote this element by $\mathbf{1}$.

Proof of assertion 19.2. For our purposes, it is enough to prove that the element $\boldsymbol{a}$ satisfies a quadratic equation

$$\boldsymbol{a}^2 + s\boldsymbol{a} + t\mathbf{1} = \mathbf{0} \tag{19.2}$$

with negative discriminant.[9]

Let n be the dimension of the algebra. Consider the $n+1$ powers of $\boldsymbol{a}$,

$$\boldsymbol{a}^0 = \mathbf{1}, \boldsymbol{a}^1, \boldsymbol{a}^2, \boldsymbol{a}^3, \ldots, \boldsymbol{a}^n.$$

In view of Theorem 9.2, this system of $n+1$ vectors is linearly dependent, so that some power must by a linear combination of its predecessors:

$$\boldsymbol{a}^m = k_{m-1}\boldsymbol{a}^{m-1} + \ldots + k_2\boldsymbol{a}^2 + k_1\boldsymbol{a} + k_0\mathbf{1}.$$

In other words, $\boldsymbol{a}$ is a root of the m-th degree equation

$$\boldsymbol{x}^m - k_{m-1}\boldsymbol{x}^{m-1} - \ldots - k_2\boldsymbol{x}^2 - k_1\boldsymbol{x} - k_0\mathbf{1} = \mathbf{0}.$$

Consider the general m-th degree polynomial

$$P(x) = x^m - k_{m-1}x^{m-1} - \ldots - k_2x^2 - k_1x - k_0.$$

Such a polynomial can be written as a product

$$P(x) = P_1(x)P_2(x)\ldots P_s(x), \tag{19.3}$$

of linear and irreducible (that is, not further decomposable) quadratic polynomials.

To follow the rest of the argument we must have a clear understanding of equality (19.3). Thus each of the polynomials $P_1(x), \ldots, P_s(x)$ is a sum of two or three terms:

$$x + t \text{ or } x^2 + sx + t.$$

The equality (19.3) states that if we multiply the polynomials $P_1(x)$, $\ldots$, $P_s(x)$ using the rule of multiplication of sums, use the formula

$$x^k \cdot x^l = x^{k+l}$$

and, finally, reduce like terms, then we obtain $P(x)$.

Note that the rules of working with powers of $\boldsymbol{a}$ are the same as the rules of working with the powers of the unknown x, that is,

$$\boldsymbol{a}^k \cdot \boldsymbol{a}^l = \boldsymbol{a}^{k+l}$$

(bear in mind the associativity of the algebra $\mathcal{A}$). It follows that equality (19.3) holds if we replace x with $\boldsymbol{a}$, that is,

$$P(\boldsymbol{a}) = P_1(\boldsymbol{a})P_2(\boldsymbol{a}) \ldots P_s(\boldsymbol{a}).$$

Since $P(\boldsymbol{a}) = \mathbf{o}$, it follows that

$$P_1(\boldsymbol{a})P_2(\boldsymbol{a}) \ldots P_s(\boldsymbol{a}) = \mathbf{o}. \tag{19.4}$$

Now we make use of the fact that $\mathcal{A}$ is a division algebra. This implies that if the product of elements is zero, then at least one of them is zero (if $\boldsymbol{uv} = \mathbf{o}$ and $\boldsymbol{u} \neq \mathbf{o}$ then, in view of the uniqueness of the solution of the equation $\boldsymbol{ux} = \mathbf{o}$, we must have $\boldsymbol{v} = \mathbf{o}$). Applied to (19.4), this means that for some i

$$P_i(\boldsymbol{a}) = \mathbf{o},$$

that is, the element $\boldsymbol{a}$ satisfies a linear or quadratic equation. If $\boldsymbol{a}$ satisfied a linear equation

$$\boldsymbol{a} + t\mathbf{1} = \mathbf{o}$$

then, contrary to the assumption, it would be proportional to $\mathbf{1}$. It follows that $\boldsymbol{a}$ satisfies an irreducible quadratic equation (19.2). Since the polynomial $P_i(x)$ is irreducible, its discriminant is negative. This proves assertion 19.2.

Proof of assertion 19.3. In the subalgebra $\mathcal{C}_{a_1}$ we choose an element $\boldsymbol{b}_1$ such that $\boldsymbol{b}_1^2 = -\mathbf{1}$ ($\boldsymbol{b}_1$ is the "imaginary unit" in the complex algebra $\mathcal{C}_{a_1}$). Similarly, in the subalgebra $\mathcal{C}_{a_2}$ we choose an element $\boldsymbol{b}_2$ such that $\boldsymbol{b}_2^2 = -\mathbf{1}$. Since $\boldsymbol{b}_1, \boldsymbol{b}_2$ differ, respectively, from $\boldsymbol{a}_1, \boldsymbol{a}_2$ by multiples of $\mathbf{1}$, it follows that the set of elements of the form $\alpha\mathbf{1} + \beta\boldsymbol{a}_1 + \gamma\boldsymbol{a}_2 + \delta\boldsymbol{a}_1\boldsymbol{a}_2$ coincides with the set of elements of the form $\alpha'\mathbf{1} + \beta'\boldsymbol{b}_1 + \gamma'\boldsymbol{b}_2 + \delta'\boldsymbol{b}_1\boldsymbol{b}_2$ that is, $\mathcal{Q}_{a_1,a_2}$ coincides with $\mathcal{Q}_{b_1,b_2}$.

Further, it is not difficult to see that if

$$\boldsymbol{e}_1 = \boldsymbol{b}_1, \quad \boldsymbol{e}_2 = k_1\boldsymbol{b}_1 + k_2\boldsymbol{b}_2,$$

and $k_2 \neq 0$, then the set $\mathcal{Q}_{e_1,e_2}$ coincides with $\mathcal{Q}_{b_1,b_2}$, and thus with $\mathcal{Q}_{a_1,a_2}$. We shall show that *it is possible to choose the numbers* k_1 *and* k_2 *so that*

$$\boldsymbol{e}_1^2 = -\mathbf{1}, \quad \boldsymbol{e}_2^2 = -\mathbf{1}, \quad (\boldsymbol{e}_1\boldsymbol{e}_2)^2 = -\mathbf{1} \tag{19.5}$$

(the first of these equalities holds for arbitrary k_1, k_2).

Note that, on the one hand,

$$(\boldsymbol{b}_1 + \boldsymbol{b}_2)^2 = \boldsymbol{b}_1^2 + \boldsymbol{b}_2^2 + (\boldsymbol{b}_1\boldsymbol{b}_2 + \boldsymbol{b}_2\boldsymbol{b}_1) = -2 \cdot \mathbf{1} + (\boldsymbol{b}_1\boldsymbol{b}_2 + \boldsymbol{b}_2\boldsymbol{b}_1),$$

and, on the other hand, the square of $\boldsymbol{b}_1 + \boldsymbol{b}_2$ must be a linear combination of $\mathbf{1}$ and $\boldsymbol{b}_1 + \boldsymbol{b}_2$:

$$(\boldsymbol{b}_1 + \boldsymbol{b}_2)^2 = p\mathbf{1} + q(\boldsymbol{b}_1 + \boldsymbol{b}_2).$$

Therefore,

$$\boldsymbol{b}_1\boldsymbol{b}_2 + \boldsymbol{b}_2\boldsymbol{b}_1 = (p+2)\mathbf{1} + q(\boldsymbol{b}_1 + \boldsymbol{b}_2). \tag{19.6}$$

Similarly,

$$(\boldsymbol{b}_1 + 2\boldsymbol{b}_2)^2 = \boldsymbol{b}_1^2 + 4\boldsymbol{b}_2^2 + 2(\boldsymbol{b}_1\boldsymbol{b}_2 + \boldsymbol{b}_2\boldsymbol{b}_1) = -5 \cdot \mathbf{1} + 2(\boldsymbol{b}_1\boldsymbol{b}_2 + \boldsymbol{b}_2\boldsymbol{b}_1),$$

and

$$(\boldsymbol{b}_1 + 2\boldsymbol{b}_2)^2 = p'\mathbf{1} + q'(\boldsymbol{b}_1 + 2\boldsymbol{b}_2);$$

so that

$$\boldsymbol{b}_1\boldsymbol{b}_2 + \boldsymbol{b}_2\boldsymbol{b}_1 = \frac{1}{2}(p'+5)\mathbf{1} + \frac{1}{2}q'(\boldsymbol{b}_1 + 2\boldsymbol{b}_2).$$

Suppose that $q \neq 0$. By equating the two expressions we could deduce that $\boldsymbol{b}_1$ differs from $\boldsymbol{b}_2$ by a multiple of $\mathbf{1}$, that is, $\boldsymbol{b}_2 \in \mathcal{C}_{b_1}$. But this is ruled out by assumption. Hence $q = 0$ and equality (19.6) implies that

$$\boldsymbol{b}_1\boldsymbol{b}_2 + \boldsymbol{b}_2\boldsymbol{b}_1 = \lambda\mathbf{1}. \tag{19.1}$$

In other words, *if* $\boldsymbol{b}_1$ *and* $\boldsymbol{b}_2$ *are two elements whose squares are* $-\mathbf{1}$, *then equality* (19.1) *holds.*

By now it is easy to determine the required elements $\boldsymbol{e}_1$ and $\boldsymbol{e}_2$. To this end we consider the element $\boldsymbol{c} = \lambda\boldsymbol{b}_1 + 2\boldsymbol{b}_2$, where λ has the same value as in (19.1). Its square is

$$\boldsymbol{c}^2 = -\lambda^2\mathbf{1} - 4 \cdot \mathbf{1} + 2\lambda(\boldsymbol{b}_1\boldsymbol{b}_2 + \boldsymbol{b}_2\boldsymbol{b}_1) = (\lambda^2 - 4) \cdot \mathbf{1}, \tag{19.7}$$

which means that $\lambda^2 - 4 < 0$.[10] Put

$$\boldsymbol{e}_2 = \frac{1}{\sqrt{4-\lambda^2}}\boldsymbol{c}.$$

Then (19.7) implies $\boldsymbol{e}_2^2 = -\mathbf{1}$, that is, the second equality in (19.5). To prove the third equality in (19.5) note that

$$\boldsymbol{e}_1\boldsymbol{e}_2 + \boldsymbol{e}_2\boldsymbol{e}_1 = \mathbf{0}. \tag{19.8}$$

Indeed,

$$\begin{aligned}\boldsymbol{e}_1\boldsymbol{e}_2 + \boldsymbol{e}_2\boldsymbol{e}_1 &= \frac{1}{\sqrt{4-\lambda^2}}(\boldsymbol{b}_1(\lambda\boldsymbol{b}_1+2\boldsymbol{b}_2) + (\lambda\boldsymbol{b}_1+2\boldsymbol{b}_2)\boldsymbol{b}_1) \\ &= \frac{1}{\sqrt{4-\lambda^2}}(-2\lambda\cdot\mathbf{1} + 2(\boldsymbol{b}_1\boldsymbol{b}_2 + \boldsymbol{b}_2\boldsymbol{b}_1)) = \mathbf{0}.\end{aligned}$$

Using (19.8), we obtain

$$(\boldsymbol{e}_1\boldsymbol{e}_2)^2 = (\boldsymbol{e}_1\boldsymbol{e}_2)(\boldsymbol{e}_1\boldsymbol{e}_2) = (\boldsymbol{e}_1\boldsymbol{e}_2)(-\boldsymbol{e}_2\boldsymbol{e}_1) = -(\boldsymbol{e}_1\boldsymbol{e}_2^2)\boldsymbol{e}_1 = \boldsymbol{e}_1^2 = -\mathbf{1}. \tag{19.9}$$

This establishes the third equality in (19.5).

Now we show that the set of elements $\mathcal{Q}_{e_1,e_2}$ of the form

$$\alpha\mathbf{1} + \beta\boldsymbol{e}_1 + \gamma\boldsymbol{e}_2 + \delta\boldsymbol{e}_1\boldsymbol{e}_2$$

(which, as mentioned earlier, coincides with $\mathcal{Q}_{a_1,a_2}$) is a subalgebra of the algebra $\mathcal{A}$.

For this it suffices to show that the product of any two of the four elements

$$\mathbf{1},\quad \boldsymbol{e}_1,\quad \boldsymbol{e}_2,\quad \boldsymbol{e}_1\boldsymbol{e}_2 \tag{19.10}$$

is itself a linear combination of these elements. The only products for which this must still be verified are the products

$$\boldsymbol{e}_1(\boldsymbol{e}_1\boldsymbol{e}_2),\quad (\boldsymbol{e}_1\boldsymbol{e}_2)\boldsymbol{e}_1,\quad \boldsymbol{e}_2(\boldsymbol{e}_1\boldsymbol{e}_2),\quad (\boldsymbol{e}_1\boldsymbol{e}_2)\boldsymbol{e}_2.$$

We have

$$\begin{aligned}\boldsymbol{e}_1(\boldsymbol{e}_1\boldsymbol{e}_2) &= \boldsymbol{e}_1^2\boldsymbol{e}_2 = -\boldsymbol{e}_2, \\ (\boldsymbol{e}_1\boldsymbol{e}_2)\boldsymbol{e}_1 &= -(\boldsymbol{e}_2\boldsymbol{e}_1)\boldsymbol{e}_1 = -\boldsymbol{e}_2\boldsymbol{e}_1^2 = \boldsymbol{e}_2, \\ \boldsymbol{e}_2(\boldsymbol{e}_1\boldsymbol{e}_2) &= -\boldsymbol{e}_2(\boldsymbol{e}_2\boldsymbol{e}_1) = -\boldsymbol{e}_2^2\boldsymbol{e}_1 = \boldsymbol{e}_1, \\ (\boldsymbol{e}_1\boldsymbol{e}_2)\boldsymbol{e}_2 &= \boldsymbol{e}_1\boldsymbol{e}_2^2 = -\boldsymbol{e}_1.\end{aligned} \tag{19.11}$$

This completes the proof of the assertion that $\mathcal{Q}_{e_1,e_2}$ is a subalgebra.

It remains to show that this subalgebra is isomorphic to the quaternion algebra. For this we show, firstly, that the four elements in (19.10) are a basis of the subalgebra in question and, secondly, that the multiplication table for this basis is the same as the multiplication table for the basis $\mathbf{1}, \boldsymbol{i}, \boldsymbol{j}, \boldsymbol{k}$ of the quaternion algebra.

By now we know that every element of $\mathcal{Q}_{e_1,e_2}$ is a linear combination of the elements in (19.10). To prove that these elements form a basis it remains to show that they are linearly independent, or (see section 8.3) that none of them is a linear combination of its predecessors. That $\boldsymbol{e}_2$ is not a linear combination of $\mathbf{1}$ and $\boldsymbol{e}_1$ follows from the fact that $\boldsymbol{e}_1$ and $\boldsymbol{e}_2$ do not belong to a single subalgebra $\mathcal{C}_{\boldsymbol{a}}$. Thus it remains to show that $\boldsymbol{e}_1\boldsymbol{e}_2$ is not a linear combination of $\mathbf{1}, \boldsymbol{e}$, and $\boldsymbol{e}_2$, that is, that we cannot have

$$\boldsymbol{e}_1\boldsymbol{e}_2 = p\boldsymbol{e}_2 + q\boldsymbol{e}_1 + r\mathbf{1}. \tag{19.12}$$

Suppose that such an equality holds. Then p and q must be different from zero (if, say, $p = 0$ then, multiplying (19.12) by $\boldsymbol{e}_1$ on the left, we would obtain the inadmissible result that $\boldsymbol{e}_2$ is a linear combination of $\mathbf{1}$ and $\boldsymbol{e}_1$). Multiplying (19.12) on the left by $\boldsymbol{e}_1$ we obtain

$$-\boldsymbol{e}_2 = p\boldsymbol{e}_1\boldsymbol{e}_2 - q\mathbf{1} + r\boldsymbol{e}_1,$$

or

$$\boldsymbol{e}_1\boldsymbol{e}_2 = -\frac{1}{p}\boldsymbol{e}_2 - \frac{r}{p}\boldsymbol{e}_1 + \frac{q}{p}\mathbf{1}.$$

The difference of the two expressions for $\boldsymbol{e}_1\boldsymbol{e}_2$ yields the equality

$$(p + \frac{1}{p})\boldsymbol{e}_2 + (q + \frac{r}{p})\boldsymbol{e}_1 + (r - \frac{q}{p})\mathbf{1} = \mathbf{0}.$$

Here the coefficient of $\boldsymbol{e}_2$ must be zero (for otherwise $\boldsymbol{e}_2$ would be a linear combination of $\mathbf{1}$ and $\boldsymbol{e}_1$), but this is impossible regardless of the choice of the real number p.

We have shown that the elements

$$\mathbf{1}, \boldsymbol{e}_1, \boldsymbol{e}_2, \boldsymbol{e}_3,$$

where $\boldsymbol{e}_3 = \boldsymbol{e}_1\boldsymbol{e}_2$, form a basis of the subalgebra $\mathcal{Q}_{e_1,e_2}$.

At this point, proving the isomorphism of the subalgebra $\mathcal{Q}_{e_1,e_2}$ and the quaternion algebra $\mathcal{Q}$ requires just one thing, namely, showing that the multiplication table for the algebra $\mathcal{Q}_{e_1,e_2}$ with the basis

$$\mathbf{1}, \boldsymbol{e}_1, \boldsymbol{e}_2, \boldsymbol{e}_3$$

is the same as the multiplication table for the quaternion algebra $\mathcal{Q}$ with basis

$$\mathbf{1}, \boldsymbol{i}, \boldsymbol{j}, \boldsymbol{k}.$$

But this follows directly from the relations (19.5), (19.8), and (19.11).

19.5 Proof of the Generalized Frobenius Theorem Based on Hurwitz's Theorem

▷ We begin with a remark bearing on the definition of an alternative algebra. We define such an algebra as one in which the following identities hold:

$$(\boldsymbol{ab})\boldsymbol{b} = \boldsymbol{a}(\boldsymbol{bb}) \text{ and } \boldsymbol{b}(\boldsymbol{ba}) = (\boldsymbol{bb})\boldsymbol{a}.$$

But there is also a second definition of an alternative algebra, according to which an algebra $\mathcal{A}$ is alternative if the value of any finite product of any two of its elements $\boldsymbol{a}$ and $\boldsymbol{b}$ does not depend on the location of parentheses in that product. This means that, for example,

$$\begin{aligned} (\boldsymbol{ab})\boldsymbol{b} &= \boldsymbol{a}(\boldsymbol{bb}), \\ (\boldsymbol{ab})(\boldsymbol{ba}) &= (\boldsymbol{a}(\boldsymbol{bb}))\boldsymbol{a}, \end{aligned}$$

and so on.

It is clear that the second definition of the alternative property implies the first. That the first definition implies the second is the content of Artin's theorem, which we shall not prove here.

In our proof of the generalized Frobenius theorem we shall use the second definition of the alternative property. This means that, strictly speaking, we shall prove the following theorem: *If a division algebra $\mathcal{A}$ has the property that any finite product of any two of its elements does not depend on the distribution of parentheses in that product, then the algebra $\mathcal{A}$ is isomorphic to one of the following four algebras: the real numbers, the complex numbers, the quaternions, and the Cayley numbers.*

It is important to note that the properties 19.1, 19.2, 19.3 of an associative division algebra hold for an alternative division algebra.

There is no need to make the slightest modification in the proofs of assertions 19.2 and 19.3. In fact, careful scrutiny of these proofs shows that we used the associativity of the algebra just twice, namely, in connection with the formula $\boldsymbol{a}^n \cdot \boldsymbol{a}^m = \boldsymbol{a}^{n+m}$, and in connection with the relation $(e_1 e_2)(e_2 e_1) = (e_1 e_2^2)e_1$, applied in the chain of equalities (19.9). Clearly, both of these relations hold in an alternative algebra.

The proof of assertion 19.1, however, must be slightly modified in the alternative case. Thus, let $\boldsymbol{e}$ be the solution of the equation $\boldsymbol{xa} = \boldsymbol{a}$. Multiplying the equality $\boldsymbol{ea} = \boldsymbol{a}$ by $\boldsymbol{e}$ on the left we obtain $\boldsymbol{e}(\boldsymbol{ea}) = \boldsymbol{ea}$, or, by the alternative property, $(\boldsymbol{ee})\boldsymbol{a} = \boldsymbol{ea}$. Hence $\boldsymbol{ee} = \boldsymbol{e}$. Again using the alternative property we have $(\boldsymbol{be})\boldsymbol{e} = \boldsymbol{b}(\boldsymbol{ee})$ and $\boldsymbol{e}(\boldsymbol{ec}) = (\boldsymbol{ee})\boldsymbol{c}$, that is,

$(\boldsymbol{be})\boldsymbol{e} = \boldsymbol{be}$ and $\boldsymbol{e}(\boldsymbol{ec}) = \boldsymbol{ec}$. It follows that $\boldsymbol{be} = \boldsymbol{b}$ and $\boldsymbol{ec} = \boldsymbol{c}$, so that $\boldsymbol{e}$ is the identity of our algebra.

To prove the generalized Frobenius theorem we could now follow the pattern of the proof of Frobenius' theorem, that is, show that if the subalgebra $\mathcal{Q}_{a,b}$ is not all of $\mathcal{A}$, then the latter contains a subalgebra isomorphic to the algebra of Cayley numbers. Then it would be necessary to prove that the latter subalgebra coincides with $\mathcal{A}$. While possible, such a proof is rather long. Therefore we shall use a different approach, namely, we shall prove that $\mathcal{A}$ is a normed algebra. By Hurwitz's theorem, this will imply the required result.

We define in the algebra $\mathcal{A}$ the following conjugation operation. If an element $\boldsymbol{a}$ is proportional to $\mathbf{1}$, then we put $\bar{\boldsymbol{a}} = a$. If $\boldsymbol{a}$ is not proportional to $\mathbf{1}$ then, by assertion 19.2, it is contained in the subalgebra $\mathcal{C}_a$. $\mathcal{C}_a$ contains a conjugate $\bar{\boldsymbol{a}}$ of $\boldsymbol{a}$, and we shall call it the conjugate of $\boldsymbol{a}$ in the algebra $\mathcal{A}$.

It follows from the definition of $\bar{\boldsymbol{a}}$ that $\bar{\bar{\boldsymbol{a}}} = \boldsymbol{a}$, and that

$$\overline{k\boldsymbol{a}} = k\bar{\boldsymbol{a}} \tag{19.13}$$

for any real number k.

Before we can deduce further properties of our conjugation we must clarify a certain issue. Suppose that an element $\boldsymbol{a}$ is not proportional to $\mathbf{1}$. Take any quaternion subalgebra $\mathcal{Q}_{a_1,a_2}$ containing $\boldsymbol{a}$. This subalgebra contains a conjugate element $\tilde{\boldsymbol{a}}$ of $\boldsymbol{a}$. The natural question is whether $\tilde{\boldsymbol{a}} = \bar{\boldsymbol{a}}$. We shall show that this is so.

As conjugates in a complex algebra, $\boldsymbol{a}$ and $\bar{\boldsymbol{a}}$ have the properties

$$\boldsymbol{a}+\bar{\boldsymbol{a}} = \text{(real number)} \cdot \mathbf{1}, \tag{19.14}$$

and

$$\boldsymbol{a}\bar{\boldsymbol{a}} = \text{(real number)} \cdot \mathbf{1}. \tag{19.15}$$

As conjugates in a quaternion algebra, $\boldsymbol{a}$ and $\tilde{\boldsymbol{a}}$ have the analogous properties

$$\boldsymbol{a}+\tilde{\boldsymbol{a}} = \text{(real number)} \cdot \mathbf{1}, \tag{19.14$'$}$$

and

$$\boldsymbol{a}\tilde{\boldsymbol{a}} = \text{(real number)} \cdot \mathbf{1}. \tag{19.15$'$}$$

Forming the differences of the equalities (19.14) and (19.14′) and (19.15) and (19.15′), we obtain the equalities

$$\bar{\boldsymbol{a}} - \tilde{\boldsymbol{a}} = \text{(real number)} \cdot \mathbf{1},$$

and

$$\boldsymbol{a}(\bar{a} - \tilde{a}) = \text{(real number)} \cdot \mathbf{1}.$$

If we had $\bar{a} \neq \tilde{a}$, then the latter relations would imply that $\boldsymbol{a}$ is a multiple of $\mathbf{1}$—an outcome that contradicts our assumption.

Thus *the conjugation of an element* $\boldsymbol{a}$ *is the same regardless of whether we think of it as an element of a complex subalgebra* $\mathcal{C}_a$ (that is, as a complex number) *or an element of a quaternion subalgebra* $\mathcal{Q}_{a_1,a_2}$ (that is, as a quaternion).

Incidentally, the same is true of the absolute value of $\boldsymbol{a}$; since (absolute value of $\boldsymbol{a})^2 = \boldsymbol{a}\bar{\boldsymbol{a}}$ in the case of complex numbers as well as quaternions, the absolute value of $\boldsymbol{a}$ is the same regardless of whether we think of it as an element of a complex subalgebra or an element of a quaternion subalgebra.

From what has just been proved about the properties of conjugation it is easy to deduce the following equalities for any two elements $\boldsymbol{a}$ and $\boldsymbol{b}$ in the algebra $\mathcal{A}$:

$$\overline{\boldsymbol{a}+\boldsymbol{b}} = \bar{\boldsymbol{a}} + \bar{\boldsymbol{b}}, \tag{19.16}$$

$$\overline{\boldsymbol{a}\boldsymbol{b}} = \bar{\boldsymbol{b}}\bar{\boldsymbol{a}}. \tag{19.17}$$

In fact, if $\boldsymbol{a}$ and $\boldsymbol{b}$ are in the same complex subalgebra (that is, if $\mathcal{C}_a$ and $\mathcal{C}_b$ coincide), then the above equalities express properties of conjugation in that subalgebra. If $\boldsymbol{b}$ is not in $\mathcal{C}_a$, than these equalities still hold for, in this case, they express properties of conjugation in $\mathcal{Q}_{a,b}$.

Formula (19.17) and $\bar{\bar{\boldsymbol{b}}} = \boldsymbol{b}$ imply that the conjugate of $\boldsymbol{a}\bar{\boldsymbol{b}}$ is $\boldsymbol{b}\bar{\boldsymbol{a}}$. It follows that

$$\boldsymbol{a}\bar{\boldsymbol{b}} + \boldsymbol{b}\bar{\boldsymbol{a}} = \text{ real number } \cdot \mathbf{1}.$$

We define in the algebra $\mathcal{A}$ a scalar product $(\boldsymbol{a}, \boldsymbol{b})$ by means of the formula

$$\boldsymbol{a}\bar{\boldsymbol{b}} + \boldsymbol{b}\bar{\boldsymbol{a}} = 2(\boldsymbol{a}, \boldsymbol{b}) \cdot \mathbf{1}.$$

The properties of a scalar product are

(1) $$(\boldsymbol{a}, \boldsymbol{a}) > 0 \text{ if } \boldsymbol{a} \neq \mathbf{o}, \text{ and } (\mathbf{o}, \mathbf{o}) = 0;$$

(2) $$(\boldsymbol{a}, \boldsymbol{b}) = (\boldsymbol{b}, \boldsymbol{a});$$

(3) $$(\boldsymbol{a}, k\boldsymbol{b}) = k(\boldsymbol{a}, \boldsymbol{b});$$

(4) $$(\boldsymbol{a}, \boldsymbol{b}_1 + \boldsymbol{b}_2) = (\boldsymbol{a}, \boldsymbol{b}_1) + (\boldsymbol{a}, \boldsymbol{b}_2),$$

and it is easy to verify that the operation we've just defined satisfies all of them. In fact, it is obvious that our operation has property (2). Formulas (19.13) and (19.16) imply that it has properties (3) and (4). To see that it has property (1) it suffices to note that

$$(\boldsymbol{a}, \boldsymbol{a}) \cdot \mathbf{1} = \boldsymbol{a}\bar{\boldsymbol{a}} = \text{ (absolute value of } \boldsymbol{a})^2 \cdot \mathbf{1}, \tag{19.18}$$

and to recall that the absolute value of a complex number is strictly positive if $\boldsymbol{a} \neq \mathbf{o}$ and zero if $\boldsymbol{a} = \mathbf{o}$.

We note that equality (19.18) implies that

$$\sqrt{(\boldsymbol{a}, \boldsymbol{a})} = \text{ absolute value of } \boldsymbol{a},$$

that is, that the norm of an element $\boldsymbol{a}$ in the algebra $\mathcal{A}$ coincides with the absolute value of $\boldsymbol{a}$ viewed as a complex number (or a quaternion).

Since any two elements $\boldsymbol{a}, \boldsymbol{b}$ in the algebra $\mathcal{A}$ belong to a single complex or quaternion subalgebra, it follows that

$$(\text{ absolute value of } \boldsymbol{ab})^2 = (\text{ absolute value of } \boldsymbol{a})^2 \cdot (\text{ absolute value of } \boldsymbol{b})^2$$

(for the algebra of complex numbers and the quaternion algebra are both normed), or that

$$(\boldsymbol{ab}, \boldsymbol{ab}) = (\boldsymbol{a}, \boldsymbol{a})(\boldsymbol{b}, \boldsymbol{b}).$$

But this equality states that $\mathcal{A}$ is a normed algebra. According to Hurwitz's theorem, the algebra $\mathcal{A}$ must be isomorphic to one of the four "standard" algebras of real numbers, complex numbers, quaternions, and Cayley numbers. This completes the proof of the generalized Frobenius theorem. $\lhd$

Chapter 20

Commutative Division Algebras

20.1 Formulation of the Main Result

In the previous chapter we found all associative division algebras. Below we describe all *commutative* division algebras.

First we state without proof the following fact: *The dimension of a commutative division does not exceed two.*[11]

It follows that in order to solve our problem we must7find all 2-dimensional division algebras.

To formulate the answer to this problem we introduce the symbol $A(\alpha, \beta, \gamma)$ to denote a 2-dimensional commutative algebra with basis $\boldsymbol{k}_1, \boldsymbol{k}_2$ and multiplication table determined by

$$\begin{aligned} \boldsymbol{k}_1 \circ \boldsymbol{k}_1 &= a\boldsymbol{k}_1 + \beta \boldsymbol{k}_2, \\ \boldsymbol{k}_2 \circ \boldsymbol{k}_2 &= -a\boldsymbol{k}_1 - \beta \boldsymbol{k}_2, \\ \boldsymbol{k}_1 \circ \boldsymbol{k}_2 &= \beta \boldsymbol{k}_1 + \gamma \boldsymbol{k}_2, \end{aligned} \tag{20.1}$$

where the numbers α, β, γ satisfy the following conditions:

(1) $$a\gamma - \beta^2 = \pm 1;$$

(2) $$\beta \geq 0;$$

(3) $$\alpha \geq 0. \text{ If } \alpha = 0, \text{ then } \gamma \geq 0.$$

The following theorem holds.

Theorem 20.1 *Every 2-dimensional commutative division algebra $\mathcal{A}$ is isomorphic to an algebra $\mathcal{A}(\alpha,\beta,\gamma)$. All algebras $\mathcal{A}(\alpha,\beta,\gamma)$ are division algebras and no two of them are isomorphic.*

The rest of this section is devoted to a proof of this theorem.

20.2 The Connection Between Multiplication in the Algebra $\mathcal{A}$ and Multiplication of Complex Numbers

Let $\mathcal{A}$ be a commutative division algebra. Denote its multiplication by $\boldsymbol{x}\Box\boldsymbol{y}$. Let $\boldsymbol{a} \neq \mathbf{o}$ be any element of $\mathcal{A}$, and consider the mapping A given by $\boldsymbol{x} \mapsto \boldsymbol{a}\Box\boldsymbol{x}$. Obviously, this mapping is linear. Since $\mathcal{A}$ is a division algebra, the mapping A has an inverse A^{-1}.

We introduce in our algebra a new multiplication given by

$$\boldsymbol{x} \cdot \boldsymbol{y} = A^{-1}(\boldsymbol{x})\Box A^{-1}(\boldsymbol{y}). \tag{20.2}$$

The algebra with the multiplication $\boldsymbol{x} \cdot \boldsymbol{y}$ is again a division algebra (the unique solvability of the equations $\boldsymbol{a}\Box\boldsymbol{x} = \boldsymbol{b}$ and $\boldsymbol{x}\Box\boldsymbol{a} = \boldsymbol{b}$ implies the unique solvability of the equations $\boldsymbol{a} \cdot \boldsymbol{x} = \boldsymbol{b}$ and $\boldsymbol{x} \cdot \boldsymbol{a} = \boldsymbol{b}$). Its identity is the element $\boldsymbol{a}\Box\boldsymbol{a}$ (for proof see section 18.2, where we consider a similar construction). But the only 2-dimensional division algebra with an identity is the algebra of complex numbers (see chapter 2). It follows that we may regard the elements $\boldsymbol{x}$ and $\boldsymbol{y}$ as complex numbers and the operation $\boldsymbol{x} \cdot \boldsymbol{y}$ as ordinary multiplication of complex numbers.

Now we denote $A^{-1}(\boldsymbol{x})$ by u and $A^{-1}(\boldsymbol{y})$ by v, and write (20.2) as

$$\boldsymbol{u}\Box\boldsymbol{v} = A(\boldsymbol{u}) \cdot A(\boldsymbol{v}).$$

This expresses the multiplication in the initial algebra $\mathcal{A}$ in terms of the multiplication of complex numbers.

As our next step, we consider the multiplication

$$\boldsymbol{u} \circ \boldsymbol{v} = A(\boldsymbol{u} \cdot \boldsymbol{v}) \tag{20.3}$$

and show that the algebras with the respective multiplications $\boldsymbol{u}\Box\boldsymbol{v}$ and $\boldsymbol{u} \circ \boldsymbol{v}$ are isomorphic.

To this end we write the multiplication table of the multiplication $\circ$ relative to some basis $\boldsymbol{e}_1, \boldsymbol{e}_2$ and show that it is the same as the multiplication table of the multiplication $\Box$ relative to the basis $\boldsymbol{e}_1' = A^{-1}(\boldsymbol{e}_1), \boldsymbol{e}_2' = A^{-1}(\boldsymbol{e}_2)$.

In fact, let

$$e_i \circ e_j = \alpha e_1 + \beta e_2.$$

Then

$$\begin{aligned} e'_i \Box e'_j &= A(e'_i) \cdot A(e'_j) = e_i \cdot e_j = A^{-1}(e_i \circ e_j) \\ &= A^{-1}(\alpha e_1 + \beta e_2) = \alpha A^{-1}(e_1) + \beta A^{-1}(e_2) = \alpha e'_1 + \beta e'_2. \end{aligned}$$

This proves the coincidence of the two multiplication tables.

We have shown that the initial algebra $\mathcal{A}$ is isomorphic to the algebra with the multiplication (20.3), where $\boldsymbol{u} \cdot \boldsymbol{v}$ is ordinary multiplication of complex numbers and A is a certain invertible linear transformation. On the other hand, it is clear that such an algebra is a commutative division algebra. Thus the problem of finding all 2-dimensional commutative division algebras reduces to the problem of finding the nonisomorphic algebras (20.3).

20.3 Determination of the Algebra $\mathcal{A}(\alpha, \beta, \gamma)$ that is Isomorphic to the Algebra $\mathcal{A}$

We must find in the algebra (20.3) a basis $\boldsymbol{k}_1, \boldsymbol{k}_2$ relative to which the multiplication table has the form (20.1) (with suitable conditions on α, β, γ). First we write down the multiplication table of the algebra (20.3) relative to the basis

$$\boldsymbol{e}_1 = \boldsymbol{1}, \quad \boldsymbol{e}_2 = \boldsymbol{i}.$$

Since

$$\begin{aligned} e_1 \circ e_1 &= A(e_1 \cdot e_1) = A(\boldsymbol{1}), \\ e_2 \circ e_2 &= A(e_2 \cdot e_2) = A(-\boldsymbol{1}) = -A(\boldsymbol{1}), \\ e_1 \circ e_2 &= A(e_1 \cdot e_2) = A(\boldsymbol{i}), \end{aligned}$$

it follows that if we put $A(\boldsymbol{1}) = a + b\boldsymbol{i}$, $A(\boldsymbol{i}) = c + d\boldsymbol{i}$, then

$$\begin{aligned} e_1 \circ e_1 &= ae_1 + be_2, \\ e_2 \circ e_2 &= -ae_1 - be_2, \\ e_1 \circ e_2 &= ce_1 + de_2. \end{aligned} \tag{20.4}$$

Now we pose the following question: Is there a basis other than $\boldsymbol{e}_1, \boldsymbol{e}_2$ (that is, $\boldsymbol{1}, \boldsymbol{i}$) relative to which the multiplication table is analogous to

(20.4):

$$\begin{aligned} \boldsymbol{k}_1 \circ \boldsymbol{k}_1 &= \alpha \boldsymbol{k}_1 + \beta \boldsymbol{k}_2, \\ \boldsymbol{k}_2 \circ \boldsymbol{k}_2 &= -\alpha \boldsymbol{k}_1 - \beta \boldsymbol{k}_2, \\ \boldsymbol{k}_1 \circ \boldsymbol{k}_2 &= \delta \boldsymbol{k}_1 + \gamma \boldsymbol{k}_2. \end{aligned} \tag{20.5}$$

In other words, are there bases in which

$$\boldsymbol{k}_1 \circ \boldsymbol{k}_1 = -\boldsymbol{k}_2 \circ \boldsymbol{k}_2. \tag{20.6}$$

Since equality (20.6) is equivalent to $A(\boldsymbol{k}_1 \cdot \boldsymbol{k}_1) = -A(\boldsymbol{k}_2 \circ \boldsymbol{k}_2)$, or $A(\boldsymbol{k}_1 \cdot \boldsymbol{k}_1) = A(-\boldsymbol{k}_2 \cdot \boldsymbol{k}_2)$, the existence of the inverse transformation A^{-1} implies that $\boldsymbol{k}_1 \circ \boldsymbol{k}_1 = -\boldsymbol{k}_2 \cdot \boldsymbol{k}_2$, or

$$\boldsymbol{k}_2 = \pm \boldsymbol{i} \boldsymbol{k}_1.$$

Hence *all bases in which the multiplication table has the form* (20.5) *are given by*

$$\boldsymbol{k}_1 = \boldsymbol{f}, \quad \boldsymbol{k}_2 = \pm \boldsymbol{i} \boldsymbol{f}, \tag{20.7}$$

where $\boldsymbol{f}$ *is an arbitrary nonzero complex number.*

Of course, there are infinitely many bases (20.7). We shall show that in one of them

$$\beta = \delta,$$

that is, in one of these bases the multiplication table has the form (20.1).

To show this we again take as the initial basis the basis $\boldsymbol{e}_1 = \boldsymbol{1}, \boldsymbol{e}_2 = \boldsymbol{i}$, and write the required basis in the form

$$\begin{aligned} \boldsymbol{k}_1 &= \rho(\cos\varphi + \boldsymbol{i}\sin\varphi), \\ \boldsymbol{k}_2 &= \pm \boldsymbol{i}\rho(\cos\varphi + \boldsymbol{i}\sin\varphi). \end{aligned}$$

To find ρ and φ we must:

1. compute the product $\boldsymbol{k}_1 \circ \boldsymbol{k}_1$ starting with the formulas (20.4) and express it as a linear combination of $\boldsymbol{k}_1$ and $\boldsymbol{k}_2$;

2. compute in a similar manner the product $\boldsymbol{k}_1 \circ \boldsymbol{k}_2$ and express it as a linear combination of $\boldsymbol{k}_1$ and $\boldsymbol{k}_2$;

3. equate the coefficient of $\boldsymbol{k}_2$ in the first of these linear combinations with the coefficient of $\boldsymbol{k}_1$ in the second.

We leave this computational task to the reader. The result is that ρ is unrestricted and φ is determined from the condition

$$\tan\varphi = \frac{b-c}{a+d}.$$

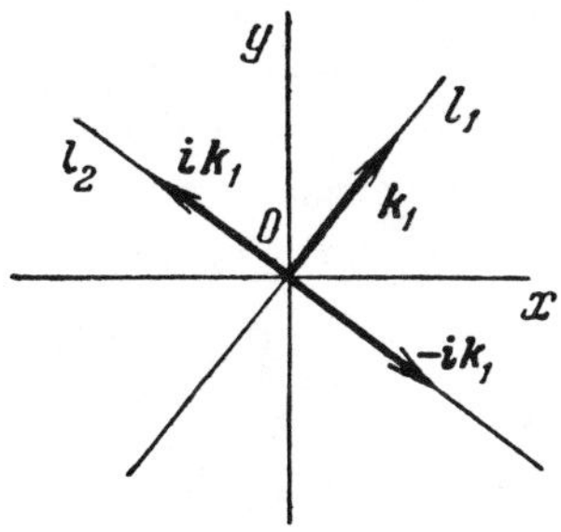

Figure 20.1.

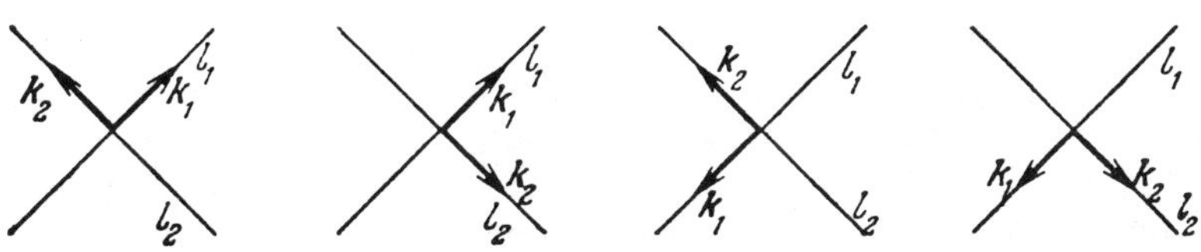

Figure 20.2.

It is clear that this condition determines φ to within a summand $n\pi$. This means that it determines in the plane two rays (Figure 20.1) that determine a single line l_1. On that line is a vector representing the complex number $\boldsymbol{k}_1$ (or, as we shall say, the vector $\boldsymbol{k}_1$). The second vector $\boldsymbol{k}_2 = \pm i\boldsymbol{k}_1$ is on its line l_2. The lengths of these vectors are the same.

Thus the totality of bases $\boldsymbol{k}_1, \boldsymbol{k}_2$ in which the multiplication table has the form (20.1) can be describe as follows. We choose a basis vector $\boldsymbol{k}_1$ on a uniquely determined line l_1 and a basis vector $\boldsymbol{k}_2$ on a line l_2 perpendicular to l_1. Both vectors must have the same length ρ. We note that for a given length there are just four of the required bases (Figure 20.2).

Upon transition from a basis $\boldsymbol{k}_1, \boldsymbol{k}_2$ to a basis $\lambda\boldsymbol{k}_1, \lambda\boldsymbol{k}_2$, where λ is a positive real number, the coefficients in the multiplication table (20.1) are multiplied by λ. Hence the additional condition

$$a\gamma - \beta^2 = \pm 1$$

determines a unique value of λ. In this way, the infinite set of admissible bases is restricted to just four (Figure 20.2).

To recapitulate: There are just four bases

$$\boldsymbol{k}_1 = \pm\boldsymbol{k}, \boldsymbol{k}_2 = \pm i\boldsymbol{k},$$

for which the multiplication table is of the form

$$\begin{array}{rcl} \boldsymbol{k}_1 \circ \boldsymbol{k}_1 & = & a\boldsymbol{k}_1 + \beta\boldsymbol{k}_2, \\ \boldsymbol{k}_2 \circ \boldsymbol{k}_2 & = & -a\boldsymbol{k}_1 - \beta\boldsymbol{k}_2, \\ \boldsymbol{k}_1 \circ \boldsymbol{k}_2 & = & \beta\boldsymbol{k}_1 + \gamma\boldsymbol{k}_2, \end{array} \tag{20.1}$$

with

$$\alpha\gamma - \beta^2 = \pm 1.$$

The last step is to show that among these four bases there is one for which $\beta \geq 0$ and $\alpha \geq 0$, and such that if $\alpha = 0$, then $\gamma \geq 0$.

In fact, if in the basis $\boldsymbol{k}_1, \boldsymbol{k}_2$ we have $\beta < 0$, then transition to the basis $\boldsymbol{k}_1, -\boldsymbol{k}_2$ yields a new table with $\beta > 0$. Similarly, if $\alpha < 0$ then, by multiplying the first basis vector by -1, we obtain a table with $\alpha > 0$ (and the sign of β unchanged). Finally, if $\alpha = 0$, then the same transformation enables us to change the sign of γ.

In sum, for every 2-dimensional commutative division algebra there exists a basis for which the multiplication table has the form (20.1) and

(1) $$\alpha\gamma - \beta^2 = \pm 1,$$

(2) $$\beta \geq 0,$$

(3) $$\alpha \geq 0 \text{ and, if } \alpha = 0, \text{ then } \gamma \geq 0.$$

This basis is, in general, unique. In certain special cases (when $\beta = 0$ or $\alpha = \gamma = 0$) there are two such bases but the table (20.1) is the same for both.

We see that with each algebra $\mathcal{A}$ there is associated a unique table (20.1) with the indicated restrictions on α, β, γ. In other words, *the algebra $\mathcal{A}$ is isomorphic to just one algebra $\mathcal{A}(\alpha, \beta, \gamma)$.*

That every algebra $\mathcal{A}(\alpha, \beta, \gamma)$ is a division algebra follows from the fact, say, that its multiplication is of the form

$$\boldsymbol{u} \circ \boldsymbol{v} = A(\boldsymbol{u} \cdot \boldsymbol{v}),$$

where the action of the transformation A is given by the formulas

$$A(\boldsymbol{k}_1) = \alpha \boldsymbol{k}_1 + \beta \boldsymbol{k}_2,$$
$$A(\boldsymbol{k}_2) = \beta \boldsymbol{k}_1 + \gamma \boldsymbol{k}_2,$$

with $\alpha\gamma - \beta^2 \neq 0$. In fact, the condition $\alpha/\beta \neq \beta/\gamma$ (equivalent to the condition $\alpha\gamma - \beta^2 \neq 0$ when $\beta\gamma \neq 0$) implies that $A(\boldsymbol{k}_1) \neq \lambda A(\boldsymbol{k}_2)$, so that the vectors $A(\boldsymbol{k}_1)$ and $A(\boldsymbol{k}_2)$ form a basis. From this it readily follows that the transformation A is invertible. In turn, this shows that the algebra with the multiplication $\boldsymbol{u} \circ \boldsymbol{v}$ is a division algebra. This completes the proof of our theorem.

Chapter 21

Conclusion

Most of the discussion in this book bears on the initial stage of the development of the theory of algebras. Now we want to touch on some newer results in this theory.

The development of the theory of algebras began with Hamilton's paper on quaternions published in 1843. Subsequently, Hamilton gave a detailed account of the results in that paper, as well as of a number of additional results, in his *Lectures on quaternions.* Hamilton's ideas were extremely influential. They prepared the ground for a series of papers on associative algebras that culminated in the proving of a number of deep theorems on the structure of such algebras.

In order to describe these theorems we must first state precisely an issue we have, so far, left out of account. It bears on the coefficients $a_1, a_2, \dots, a_n$ in the expression

$$a_1\boldsymbol{i}_1 + a_2\boldsymbol{i}_2 + \ldots + a_n\boldsymbol{i}_n \tag{21.1}$$

for the elements of an n-dimensional algebra. So far, we've always assumed that these coefficients were real numbers; in technical terms, then, we've been discussing *algebras over the field of real numbers.* But there are occasions when one discusses algebras whose elements are given by expressions of the form (21.1) in which the coefficients $a_1, a_2, \dots, a_n$ are arbitrary complex numbers. Such algebras are called *algebras over the field of complex numbers.* In addition to the fields of real and complex numbers there are many other fields[12] (for example, the field of rational numbers), and thus a corresponding multitude of other types of algebras.

Many results in the theory of algebras vary drastically with the field of the coefficients $a_1, \dots, a_n$ in the expressions (21.1), that is, the field over which the algebra is being considered. For example, we know that over the field of real numbers there are (in addition to the reals themselves) three associative division algebras (and infinitely many nonassociative division algebras), at a time when there is *just one complex division algebra,* namely, the 1-dimensional algebra of complex numbers. Apart from it, there is no additional (associative or nonassociative) division algebra over the field of complex numbers. This is easy to prove but we won't do it here.

We shall require certain definitions.

1. By an *ideal* of an algebra $\mathcal{A}$ we mean a subspace $\mathcal{U}$ such that

   $$\mathcal{A}\mathcal{U} \subset \mathcal{U} \text{ and } \mathcal{U}\mathcal{A} \subset \mathcal{U}.$$

   This means that for any elements $\boldsymbol{a} \in \mathcal{A}$ and $\boldsymbol{u} \in \mathcal{U}$ the two products $\boldsymbol{au}$ and $\boldsymbol{ua}$ are in $\mathcal{U}$. In other words, the product of an element in the ideal by an element of the algebra (in either order) is again an element of the ideal.

   Incidentally, the two extreme cases—when the subspace $\mathcal{U}$ coincides with all of $\mathcal{A}$ or consists of just the one element $\mathbf{0}$ are not regarded as ideals (they are sometimes referred to as trivial ideals).

   One example of an ideal is the subspace of elements $b\boldsymbol{\Omega}$ in the algebra of dual numbers (that is, numbers of the form $a + b\boldsymbol{\Omega}$ with $\boldsymbol{\Omega}^2 = \mathbf{0}$). Another example is the subspace of elements of the form $a(\mathbf{1} + \boldsymbol{E})$ in the algebra of double numbers (that is, numbers of the form $a + b\boldsymbol{E}$ with $\boldsymbol{E}^2 = \mathbf{1}$).

2. An algebra without nontrivial ideals is called *simple.*

   We might say that the concept of a simple algebra is a generalization of the concept of a division algebra. Every division algebra is necessarily simple. In fact, if an algebra has an ideal $\mathcal{U}$, then the equation

   $$\boldsymbol{ux} = \boldsymbol{b},$$

   where $\boldsymbol{u}$ is in $\mathcal{U}$ and $\boldsymbol{b}$ is not, is not solvable. Hence such an algebra cannot be a division algebra.

   At the end of the 19th century research in the theory of algebras centered, for the most part, on associative algebras (as noted earlier, the term "algebra" was taken to mean "associative algebra"). This produced a fairly clear understanding of the structure of associative algebras. The first substantial result dealt with simple algebras and was obtained in 1893 by Molien. Frobenius and Cartan discovered the same result independently of Molien. It turned out that, up to isomorphism, all complex simple algebras are full matrix algebras of order n (that is, algebras consisting of all square matrices of order n).

In 1907 the American mathematician Wedderburn proved a more general result for algebras over an arbitrary field $\mathcal{P}$. Wedderburn's result was to the effect that *all simple associative algebras over a field $\mathcal{P}$ are precisely the full matrix algebras with elements from an associative division algebra over $\mathcal{P}$.*

For example, according to this theorem the simple associative algebras over the field $\mathcal{R}$ of real numbers form three series:

(a) the algebras of matrices whose elements are real numbers;

(b) the algebras of matrices whose elements are complex numbers (these algebras are to be regarded as algebras over the field $\mathcal{R}$, so that the algebra of complex numbers, say, has dimension 2 and, similarly, the algebra of all complex matrices of order n has dimension $2n^2$);

(c) the algebras of matrices whose elements are quaternions (the dimension of the algebra of quaternion matrices of order n is $4n^2$).

If we bear in mind that the only complex division algebra is the algebra of complex numbers, then Wedderburn's theorem readily implies the previously mentioned theorem on complex simple algebras.

After the determination of all simple associative algebras it was found (by the same authors) that the structure of arbitrary associative algebras is determined to a large extent by the structure of simple associative algebras. A precise formulation of the latter assertion requires the introduction of additional concepts.

3. Let $\mathcal{U}_1$ and $\mathcal{U}_2$ be two algebras. Their *direct sum* is the new algebra $\mathcal{A}$ whose elements are all pairs

$$(\boldsymbol{u}_1, \boldsymbol{u}_2) \ (\boldsymbol{u}_1 \in \mathcal{U}_1, \boldsymbol{u}_2 \in \mathcal{U}_2)$$

and whose rules of addition and multiplication are given by

$$\begin{aligned}(\boldsymbol{u}_1, \boldsymbol{u}_2) + (\boldsymbol{u}'_1, \boldsymbol{u}'_2) &= (\boldsymbol{u}_1 + \boldsymbol{u}'_1, \boldsymbol{u}_2 + \boldsymbol{u}'_2),\\ (\boldsymbol{u}_1, \boldsymbol{u}_2) \cdot (\boldsymbol{u}'_1, \boldsymbol{u}'_2) &= (\boldsymbol{u}_1 \boldsymbol{u}'_1, \boldsymbol{u}_2 \boldsymbol{u}'_2).\end{aligned}$$

It is easy to see that the elements of the form $(\boldsymbol{u}_1, \mathbf{o})$ form a subalgebra of the algebra $\mathcal{A}$ isomorphic to $\mathcal{U}_1$; we denote it by $\mathcal{A}_1$. Similarly, the elements of the form $(\mathbf{o}, \boldsymbol{u}_2)$ form a subalgebra $\mathcal{A}_2$ isomorphic to $\mathcal{U}_2$. Both of these subalgebras are ideals. For example, the equalities

$$(\boldsymbol{u}_1, \mathbf{o})(\boldsymbol{u}'_1, \boldsymbol{u}'_2) = (\boldsymbol{u}_1 \boldsymbol{u}'_1, \mathbf{o}), \quad (\boldsymbol{u}'_1, \boldsymbol{u}'_2)(\boldsymbol{u}_1, \mathbf{o}) = (\boldsymbol{u}'_1 \boldsymbol{u}_1, \mathbf{o})$$

show that $\mathcal{A}_1$ is an ideal.

We note that the subalgebras $\mathcal{A}_1$ and $\mathcal{A}_2$ are *complementary.* This means of any number of algebras is defined similarly. The elements of the direct sum of algebras $\mathcal{U}_1, \mathcal{U}_2, \ldots, \mathcal{U}_k$ are all k-tuples

$$(\boldsymbol{u}_1, \boldsymbol{u}_2, \ldots, \boldsymbol{u}_k)(\boldsymbol{u}_1 \in \mathcal{U}_1, \ldots, \boldsymbol{u}_k \in \mathcal{U}_k).$$

An example of a direct sum is the algebra of matrices of order $p+q$ with "block-diagonal" structure

$$\left(\begin{array}{c|c} A & 0 \\ \hline 0 & B \end{array}\right).$$

Here A and B are arbitrary matrices of order p and q, respectively. It is easy to verify that the multiplication of these matrices satisfies the rule

$$\left(\begin{array}{c|c} A & 0 \\ \hline 0 & B \end{array}\right)\left(\begin{array}{c|c} A' & 0 \\ \hline 0 & B' \end{array}\right) = \left(\begin{array}{c|c} AA' & 0 \\ \hline 0 & BB' \end{array}\right),$$

which shows that this algebra is isomorphic to the direct sum of the algebra of all matrices of order p and the algebra of all matrices of order q.

It is a curious fact the algebra $\mathcal{A}$ of dual numbers considered in the early part of this book (see chapter 2) is isomorphic to the direct sum of two algebras of real numbers. In fact, take the following basis of the algebra $\mathcal{A}$

$$\boldsymbol{i}_1 = (1 - \boldsymbol{E})/2, \quad \boldsymbol{i}_2 = (1 + \boldsymbol{E})/2.$$

Clearly,

$$\boldsymbol{i}_1^2 = \boldsymbol{i}_1, \quad \boldsymbol{i}_2^2 = \boldsymbol{i}_2, \quad \boldsymbol{i}_1\boldsymbol{i}_2 = \mathbf{o}.$$

Each element $\boldsymbol{a} \in \mathcal{A}$ can be uniquely represented as a sum

$$a_1\boldsymbol{i}_1 + a_2\boldsymbol{i}_2,$$

and the multiplication of two elements satisfies the rule

$$(a_1\boldsymbol{i}_1 + a_2\boldsymbol{i}_2)(b_1\boldsymbol{i}_1 + b_2\boldsymbol{i}_2) = a_1b_1\boldsymbol{i}_1 + a_2b_2\boldsymbol{i}_2.$$

If we associate with an element $\boldsymbol{a}$ the pair of numbers (a_1, a_2), then we see that our algebra is the direct sum of two copies of the algebra $\mathcal{R}$ of real numbers.

4. A *semisimple* algebra is the direct sum of simple algebras. Since a direct sum is uniquely determined by its summands, it follows that if we know all simple associative algebras, then we know all semisimple associative algebras.

   For example, every semisimple associative algebra over the field of complex numbers is isomorphic to the algebra of all "block-diagonal" matrices with blocks of order $p_1, p_2, \ldots, p_k$ on the diagonal (the numbers $p_1, p_2, \ldots, p_k$ are fixed). In particular, for $k = 3$ we obtain matrices of the form

$$\left(\begin{array}{c|c|c} A & 0 & 0 \\ \hline 0 & B & 0 \\ \hline 0 & 0 & C \end{array}\right).$$

5. An algebra is called *nilpotent* if there is a number k such that the product of any k of its elements, with arbitrary distribution of parentheses, is zero. (Our definition of a nilpotent algebra is general in the sense that we don't assume associativity of multiplication. Hence the need for the remark about the arbitrary distribution of parentheses.)

   A subalgebra of an algebra is *nilpotent* if it is nilpotent when viewed as an algebra in its own right.

   The simplest example of a nilpotent algebra is the null algebra (the product of any two elements is zero). Another example is furnished by the algebra with basis $\boldsymbol{i}_1, \boldsymbol{i}_2, \boldsymbol{i}_3$ and multiplication table

$$\begin{cases} \boldsymbol{i}_1\boldsymbol{i}_2 = \boldsymbol{i}_3, \\ \text{the other } \boldsymbol{i}_\alpha\boldsymbol{i}_\beta \text{ are } \mathbf{o}. \end{cases}$$

   We note that the properties of nilpotent algebras are, in a sense, opposite to the properties of semisimple algebras. For example if $\mathcal{A}$ is a nilpotent algebra, a certain "power" $\mathcal{A}^k$ of $\mathcal{A}$, that is, the set of finite sums of products of elements of $\mathcal{A}$ taken k at a time, consists of zero alone. On the other hand, any power of a semisimple algebra coincides with that algebra.

   It is easy to show that if $\mathcal{V}_1$ and $\mathcal{V}_2$ are two nilpotent ideals of any algebra $\mathcal{A}$, then their sum (that is, all elements of the form $\boldsymbol{v}_1 + \boldsymbol{v}_2$, $\boldsymbol{v}_1 \in \mathcal{V}_1$, $\boldsymbol{v}_2 \in \mathcal{V}_2$) is again a nilpotent ideal. It follows readily that among the nilpotent ideals of an algebra $\mathcal{A}$ there must be a *maximal* one, that is a nilpotent ideal containing all other nilpotent ideals.

We can now formulate the fundamental theorem of the theory of associative algebras.

The Wedderburn theorem. *In any associative algebra $\mathcal{A}$ there is a semisimple subalgebra $\mathcal{U}$ complementary to its maximal nilpotent ideal $\mathcal{V}$.*

In other words, every element $\boldsymbol{a} \in \mathcal{A}$ can be uniquely represented as a sum $\boldsymbol{u} + \boldsymbol{v}$, where $\boldsymbol{u}$ belongs to a semisimple subalgebra $\mathcal{U}$ and $\boldsymbol{v}$ to the maximal nilpotent ideal $\mathcal{V}$. It follows that we can associate with every element $\boldsymbol{a}$ the pair $(\boldsymbol{u}, \boldsymbol{v})$, $\boldsymbol{u} \in \mathcal{U}$, $\boldsymbol{v} \in \mathcal{V}$. The product of any two elements of the algebra $\mathcal{A}$ satisfies the rule

$$(\boldsymbol{u}_1, \boldsymbol{v}_1)(\boldsymbol{u}_2, \boldsymbol{v}_2) = (\boldsymbol{u}_1\boldsymbol{u}_2, \boldsymbol{v}), \tag{21.2}$$

where $\boldsymbol{v} = \boldsymbol{u}_1\boldsymbol{v}_2 + \boldsymbol{v}_1\boldsymbol{u}_2 + \boldsymbol{v}_1\boldsymbol{v}_2 \in \mathcal{V}$ (one must bear in mind that $\mathcal{V}$ is an ideal).

In the special case when $\mathcal{U}$ is also an ideal, each of the products $\boldsymbol{u}_1\boldsymbol{v}_2$ and $\boldsymbol{v}_1\boldsymbol{u}_2$ is $\mathbf{o}$ (for these products are simultaneously in $\mathcal{U}$ and $\mathcal{V}$), and the multiplication rule takes the form

$$(\boldsymbol{u}_1, \boldsymbol{v}_1)(\boldsymbol{u}_2, \boldsymbol{v}_2) = (\boldsymbol{u}_1\boldsymbol{u}_2,\ \boldsymbol{v}_1\boldsymbol{v}_2).$$

In this case the algebra $\mathcal{A}$ is the direct sum of the algebras $\mathcal{U}$ and $\mathcal{V}$. In the general case, the structure of the algebra $\mathcal{A}$ is not entirely determined by the structure of the algebras $\mathcal{U}$ and $\mathcal{V}$ taken separately, for the element $\boldsymbol{v}$ in (21.2) depends not only on $\boldsymbol{v}_1, \boldsymbol{v}_2$ but also on $\boldsymbol{u}_1, \boldsymbol{u}_2$. Nevertheless, the fact that every associative algebra $\mathcal{A}$ can be represented by a set of pairs $(\boldsymbol{u}, \boldsymbol{v})$, where $\boldsymbol{u}$ ranges over a certain semisimple algebra and $\boldsymbol{v}$ over a nilpotent algebra, sheds a great deal of light on the structure of associative algebras.

To illustrate Wedderburn's theorem we consider the algebra of matrices of order $p + q$ in which the elements in the last q rows are zero. All such matrices can be written in the form

$$\left(\begin{array}{c|c} u & v \\ \hline 0 & 0 \end{array}\right), \tag{21.3}$$

where u is a square matrix of order p and v is a rectangular matrix with p rows and q columns. It is easy to show that the maximal nilpotent ideal $\mathcal{V}$ consists of the matrices (21.3) with $u = 0$ (and is a null algebra). As a complementary semisimple algebra $\mathcal{U}$ we can take the set of matrices (21.3) with $v = 0$ (in this case the subalgebra $\mathcal{U}$ is simple).

To emphasize the comprehensive nature of Wedderburn's theorem we shall give examples of 2-dimensional (necessarily nonassociative) algebras to which the theorem doesn't apply.

In the first example, the multiplication table has the form

$$\boldsymbol{e}_1\boldsymbol{e}_1 = \boldsymbol{e}_1 + \boldsymbol{e}_2, \quad \boldsymbol{e}_2\boldsymbol{e}_2 = \mathbf{o}, \quad \boldsymbol{e}_1\boldsymbol{e}_2 = \boldsymbol{e}_2, \quad \boldsymbol{e}_2\boldsymbol{e}_1 = \mathbf{o}.$$

It is easy to see that the elements of the form $\boldsymbol{ke_2}$ constitute a 1-dimensional nilpotent ideal $\mathcal{N}$. This ideal is maximal, for the only subspace that contains it is the algebra itself and that algebra is not nilpotent (no power of $\boldsymbol{e}_1$ is zero). It is easy to check that $\mathcal{A}$ contains no other subalgebras and therefore no subalgebra complementary to $\mathcal{N}$.

The second example is that of the algebra with the multiplication table

$$\boldsymbol{e}_1\boldsymbol{e}_1 = \boldsymbol{e}_1, \quad \boldsymbol{e}_2\boldsymbol{e}_2 = \boldsymbol{e}_2, \quad \boldsymbol{e}_1\boldsymbol{e}_2 = \boldsymbol{e}_2, \quad \boldsymbol{e}_2\boldsymbol{e}_1 = \mathbf{o}.$$

This algebra contains no nilpotent ideals. If this algebra "satisfied" Wedderburn's theorem, it would be simple or semisimple. The first possibility does not arise, for the algebra contains the ideal of elements of the form $\boldsymbol{ke_2}$. The second possibility does not arise because this ideal is the only ideal in our algebra.

The results obtained in the theory of associative algebras have served as a model for further investigations. Many subsequent papers were devoted to showing that Wedderburn's theorem holds for other classes of algebras (we just saw that the theorem cannot be true for all algebras) and to listing the simple algebras of these classes.

Zorn showed that Wedderburn's theorem applies to the class of alternative algebras that is larger than the class of associative algebras. Here we mention the interesting and unexpected fact that the class of alternative algebras is not much larger than that of associative algebras. In fact, in the case of the field of complex numbers, the larger class is obtained from the smaller one by the addition of the single algebra of "complex" Cayley numbers, and in the case of the field of real numbers—by the addition of a few algebras of the same type as the Cayley numbers.

We shall say a few words about two more classes of algebras for which Wedderburn's theorem is true. Start with any associative algebra $\mathcal{A}$ and use it to construct two more algebras $\mathcal{A}^+$ and $\mathcal{A}^-$, consisting of the same elements as $\mathcal{A}$ but with the following rules of multiplication:

$$\text{in } \mathcal{A}^+ \quad \boldsymbol{a} \square \boldsymbol{b} = \boldsymbol{ab} + \boldsymbol{ba},$$

$$\text{in } \mathcal{A}^- \quad \boldsymbol{a} \circ \boldsymbol{b} = \boldsymbol{ab} - \boldsymbol{ba}.$$

The algebra $\mathcal{A}^+$ is commutative and, as is easily verified,

$$(\boldsymbol{b}^2 \square \boldsymbol{a}) \square \boldsymbol{b} = \boldsymbol{b}^2 \square (\boldsymbol{a} \square \boldsymbol{b}). \tag{21.3}$$

The algebra $\mathcal{A}^-$ is anticommutative (that is, $\boldsymbol{a} \circ \boldsymbol{b} = -\boldsymbol{b} \circ \boldsymbol{a}$) and

$$\boldsymbol{a} \circ (\boldsymbol{b} \circ \boldsymbol{c}) + \boldsymbol{b} \circ (\boldsymbol{c} \circ \boldsymbol{a}) + \boldsymbol{c} \circ (\boldsymbol{a} \circ \boldsymbol{b}) = \mathbf{o}. \tag{21.4}$$

A commutative algebra in which (21.3) holds is called a *Jordan algebra* (named for the German mathematician P. Jordan). An anticommutative algebra in which (21.4) holds is called a *Lie algebra*, in honor of the Norwegian mathematician Sophus Lie who, at the end of the 19th century, was the first to investigate these algebras in connection with the theory of "continuous groups of transformations." Lie algebras play a very important role in modern mathematics and find applications in virtually every one of its areas.

The classification of Jordan algebras was carried out by the American mathematician Albert,who also proved the validity of Weddeburn's theorem for these algebras.

The fundamental theorems on the structure of Lie algebras were obtained by E. Cartan, one of the greatest mathematicians of the 20th century. In particular, Cartan classified the simple Lie groups. The extension of Wedderburn's theorem to Lie algebras is due to Levi. Here it turned out that the concept of a nilpotent ideal had to be replaced with that of a *solvable* ideal. A more detailed study of these matters is beyond the scope of the present book.

Chapter 22

Notes

1. If q_1 and q_2 are two nonzero vectors, then we say that q_1 is carried to q_2 by a rotation about (the directed line, or axis, determined by) $[q_1, q_2]$ through φ, where φ is the directed angle from q_1 to q_2, $0^0 \leq \varphi \leq 180^0$; cf. section 4.3. (Translator)

2. We assume the usual geometric interpretation of the sum of vectors and the multiplication of a vector by a number. We recall that vectors are added in accordance with the "parallelogram law," and multiplication of a vector by a number amounts to stretching it by a factor $|k|$ and, if $k < 0$, reversing the orientation of the stretched vector.

3. The word "nonzero" in the statement of the theorem is essential, for the zero vector is orthogonal to *every* vector $\boldsymbol{y}$. This follows from the equalities
$$(\mathbf{o}, \boldsymbol{y}) = (0\boldsymbol{y}, \boldsymbol{y}) = 0(\boldsymbol{y}, \boldsymbol{y}) = 0.$$

4. Here and in the rest of the book we make use of the general properties of inner products established in chapter 12.

5. The square of an element that is not orthogonal to $\mathbf{1}$, that is, an element of the form $\boldsymbol{a} = k\mathbf{1} + \boldsymbol{a}'$ with $k \neq 0$ and $\boldsymbol{a}' \perp \mathbf{1}$ is
$$(k\mathbf{1} + \boldsymbol{a}')(k\mathbf{1} + \boldsymbol{a}') = k^2\mathbf{1} + \boldsymbol{a}'^2 + 2k\boldsymbol{a}' = k^2\mathbf{1} + \mu\mathbf{1} + 2k\boldsymbol{a}'.$$
If this element were proportional to $\mathbf{1}$, then it would follow that $\boldsymbol{a}' = \mathbf{o}$, and so $\boldsymbol{a} = k\mathbf{1}$. But the square of the latter element is not equal to $\lambda\mathbf{1}$ with $\lambda \leq 0$.

6. The equality $\boldsymbol{b}' = k\mathbf{1} + \boldsymbol{b}$ implies that $(\boldsymbol{b}', \boldsymbol{b}') = k^2(\mathbf{1}, \mathbf{1}) + (\boldsymbol{b}, \boldsymbol{b})$. Also, $(\mathbf{1}, \mathbf{1}) = 1$. This follows readily from the fundamental equality (17.1) for $\boldsymbol{a} = \boldsymbol{b} = \mathbf{1}$.

7. Incidentally, the existence of the transformations A^{-1} and B^{-1} implies that each of the equations

$$\boldsymbol{x} \circ \boldsymbol{e} = \boldsymbol{a} \text{ and } \boldsymbol{e} \circ \boldsymbol{y} = \boldsymbol{a}$$

has a unique solution. The fact that $|\boldsymbol{e}| = 1$ is immaterial: what counts is that $\boldsymbol{e} \neq \mathbf{o}$. We conclude that *every normed algebra is a division algebra.*

8. In order actually to find the forms Φ_γ we must write

$$\boldsymbol{x}\boldsymbol{y} = (\sum_a x_a \boldsymbol{e}_\alpha)(\sum_\beta y_\beta \boldsymbol{k}_\beta) = \sum_{a\beta} x_a y_\beta \boldsymbol{e}_\alpha \boldsymbol{k}_\beta$$

and replace each product $\boldsymbol{e}_\alpha \boldsymbol{k}_\beta$ with a suitable linear combination of the basis vectors $\boldsymbol{i}_1, \boldsymbol{i}_2, \ldots, \boldsymbol{i}_n$.

9. In fact, (19.2) implies that $\boldsymbol{a}^2 = -s\boldsymbol{a} - t\mathbf{1}$. Hence the set of elements of the form $\alpha\mathbf{1} + \beta\boldsymbol{a}$ is closed under multiplication. As such, it is a 2-dimensional hypercomplex system. In view of section 2.2 , if $(s^2/4) - t < 0$ (the case of a negative discriminant), then the system is isomorphic to the complex numbers.

10. If we had $\rho = \lambda^2 - 4 \geq 0$, then $\boldsymbol{c}^2 = \rho\mathbf{1}$ would imply that $(\boldsymbol{c} - \sqrt{\rho} \cdot \mathbf{1})(\boldsymbol{c} + \sqrt{\rho} \cdot \mathbf{1}) = \mathbf{o}$, that is, that $\boldsymbol{c} = \sqrt{\rho} \cdot \mathbf{1}$, and $\boldsymbol{c} = -\sqrt{\rho} \cdot \mathbf{1}$. But this is impossible, for $\boldsymbol{b}_1$ and $\boldsymbol{b}_2$ do not belong to the same complex subalgebra.

11. The following ingenious proof, due to G. Špiz, relies on topological considerations. It is intended for readers familiar with elementary topological concepts.

    Let $\mathcal{A}$ be a n-dimensional commutative division algebra. If $\boldsymbol{x}$ and $\boldsymbol{y}$ are two elements such that $\boldsymbol{x}^2 = \boldsymbol{y}^2$ then, by commutativity, the product $(\boldsymbol{x} - \boldsymbol{y})(\boldsymbol{x} + \boldsymbol{y}) = \mathbf{o}$. The absence of divisors of zero implies that $\boldsymbol{x} = \boldsymbol{y}$ or $\boldsymbol{x} = -\boldsymbol{y}$. Hence the mapping $\boldsymbol{x} \mapsto \boldsymbol{x}^2$ induces a continuous monomorphism of the sphere S^{n-1} into the projective space RP^{n-1}. As is well known, such a mapping is possible only for $n = 2$.

12. A field is defined as follows. Let $\mathcal{P}$ be a set of elements with two operations called *addition* and *multiplication* denoted, respectively,

by $a+b$ and $a \cdot b$. The set $\mathcal{P}$ is called a field if both operations defined in it are commutative and associative, multiplication is distributive over addition, and it is possible to subtract (that is, the equation $a + x = b$ is uniquely solvable) and divide. The latter means that if $a \neq 0$, then the equation $ax = b$ is uniquely solvable; here 0 is the element of $\mathcal{P}$ such that $a + 0 = a$ for all $a \in \mathcal{P}$ (the existence of such an element is easily demonstrated).

In our definition, the word "operation" refers to any rule that associates with every pair of elements $a \in \mathcal{P}, b \in \mathcal{P}$ a definite third element $c \in \mathcal{P}$.